THE COMPLETE IDIOT'S GUIDE® TO

String Theory

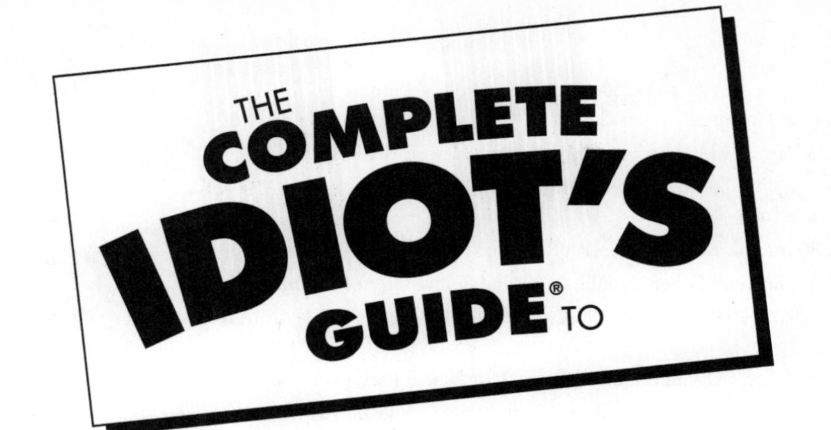

String Theory

by George Musser

A member of Penguin Group (USA) Inc.

ALPHA BOOKS

Published by the Penguin Group

Penguin Group (USA) Inc., 375 Hudson Street, New York, New York 10014, USA

Penguin Group (Canada), 90 Eglinton Avenue East, Suite 700, Toronto, Ontario M4P 2Y3, Canada (a division of Pearson Penguin Canada Inc.)

Penguin Books Ltd., 80 Strand, London WC2R 0RL, England

Penguin Ireland, 25 St. Stephen's Green, Dublin 2, Ireland (a division of Penguin Books Ltd.)

Penguin Group (Australia), 250 Camberwell Road, Camberwell, Victoria 3124, Australia (a division of Pearson Australia Group Pty. Ltd.)

Penguin Books India Pvt. Ltd., 11 Community Centre, Panchsheel Park, New Delhi—110 017, India

Penguin Group (NZ), 67 Apollo Drive, Rosedale, North Shore, Auckland 1311, New Zealand (a division of Pearson New Zealand Ltd.)

Penguin Books (South Africa) (Pty.) Ltd., 24 Sturdee Avenue, Rosebank, Johannesburg 2196, South Africa

Penguin Books Ltd., Registered Offices: 80 Strand, London WC2R 0RL, England

International Standard Book Number: 978-1-59257-702-6
Library of Congress Catalog Card Number: 2008920831

10 09 08 8 7 6 5 4 3 2

Interpretation of the printing code: The rightmost number of the first series of numbers is the year of the book's printing; the rightmost number of the second series of numbers is the number of the book's printing. For example, a printing code of 08-1 shows that the first printing occurred in 2008.

Printed in the United States of America

Publisher: *Marie Butler-Knight*
Editorial Director: *Mike Sanders*
Senior Managing Editor: *Billy Fields*
Acquisitions Editor: *Tom Stevens*
Development Editor: *Susan Zingraf*
Production Editor: *Kayla Dugger*
Copy Editor: *Nancy Wagner*
Cartoonist: *Steve Barr*
Cover Designer: *Kurt Owens*
Book Designer: *Trina Wurst*
Indexer: *Tonya Heard*
Layout: *Ayanna Lacey*
Proofreader: *John Etchison*

For my parents, George and Judith Musser

Contents at a Glance

Part 1: String Theory in a Nutshell 1

1 What Is String Theory? 3
I break the laws of comedy by giving away my punch lines about what string theory is; what the leading alternative theory, known as loop gravity, is; and where they stand.

2 Why Should Anybody Care? 13
Conceptual advances in physics have played a huge role in shaping the modern world. A unified theory of physics, such as string theory, may prove equally revolutionary.

Part 2: The Great Clash of Worldviews 25

3 Einstein's Theories 27
What commuting by New Jersey Transit trains has to say about the nature of space and time.

4 The Quantum Revolution 43
Quantum theory is no mere theory. It's a whole way of viewing the world that manages to be both compelling and counterintuitive at the same time.

5 The Standard Model of Particles 59
From a few varieties of building blocks, our whole universe emerges.

6 The World of the Small 71
No ordinary microscope can probe particles, let alone whatever makes them up. You need to study matter under ridiculously extreme conditions.

Part 3: The Need for Unity 85

7 Why Unify? 87
Like the heroes of Greek mythology, quantum theory and Einstein's theories of relativity have their tragic flaws. Fixing them may require merging them.

8 Black Holes 99
That giant sucking sound you hear is quantum theory and relativity being devoured by black holes. These cosmic sinkholes are Exhibit A for why physicists are so keen on unification.

9 The Big Bang 111
Neither relativity nor quantum theory can explain the origin of our universe on its own. So looking back in cosmic history forces us to dig deeper into the workings of nature.

10 Time Machines 125
Einstein's theories allow for time travel, so why aren't we overrun with tourists from the year 2500? Time travel strikes most physicists as impossible, but it'll take a unified theory to know for sure.

Part 4: Gravity Meets the Quantum 135

11 The Paradox of the Graviton 137
If gravity acts like other forces, particles called gravitons should transmit it. But a first crack at describing gravitons fails abjectly.

12 The Music of Strings 147
In fits and starts, string theory has become the leading effort to get relativity and quantum theory to dovetail and gravitons to behave themselves.

13 Playing a Different Tune 159
String theory is hardly the only possible approach to taming the graviton. Other contenders include loop quantum gravity, buckyspace, causal sets, and fluids-and-solids analogies.

Part 5: The Big Ideas 169

14 Extra Dimensions 171
My apologies in advance: this chapter may blow your mind. String theory says that space may have more than the three dimensions we see.

15 Parallel Universes 187
Our universe may be just one among many in a vast multiverse. This idea makes sense of many of the features of the world that otherwise have no rhyme or reason.

16 The Root of the Tree 199
What would happen if we could zoom all the way in to the tiniest distances of space? The very concept of space might dissolve.

17 Symmetry 211
What makes physicists think they're measuring something objective? One reason is that the laws of physics are symmetrical, meaning that they correct for our biased points of view and arbitrary conventions.

18 Emergence 225
What the heck are space and time, anyway? Are they built up of even deeper building blocks? String theory suggests they are.

Part 6: What Has String Theory Done for You Lately? 237

19 Black Branes and Balls of String 239
What do string theory and loop gravity have to say about the main thing that motivated them, black holes?

20 Before the Big Bang 249
Both string theory and loop gravity suggest that the start of the big bang was not the beginning of time but a transition from a pre-existing universe.

21 Ten Ways to Test String Theory 261
Ongoing and upcoming experiments might not be able to prove or disprove string theory and other such theories but will let physicists know whether they're on the right track.

22 The String Wars 277
Controversy has swirled around string theory ever since it became a viable unified theory in the early 1980s. Everyone needs to chill.

23 What Now? 287
All the candidate theories are still works in progress at best. So what's missing?

Appendixes

A Glossary 297

B Selected Readings 315

Index 325

Contents

Part 1: String Theory in a Nutshell 1

1 What Is String Theory? 3

The Ultimate Symphony 3
Alternative Music 7
Big Things Come in Small Packages 8
String Instruments 9

2 Why Should Anybody Care? 13

The Tree of Physics 13
The Joy of Unification 15
Why Is This Theory Unlike All Others? 18
Big Ideas Don't Like to Be Cooped Up 20
Sense and Transcendence 22
A Shared Effort 23

Part 2: The Great Clash of Worldviews 25

3 Einstein's Theories 27

Trains of Thought 27
There's No Time Like the Present 30
Slow Down, You Move Too Fast 32
Space and Time, Unite! 34
Think Globally, Act Locally 36
Mass and Energy 37
Atlas Shrugs 39

4 The Quantum Revolution 43

Not as Weird as They Say 43
Seven Insights of Quantum Theory 45
The Games Quanta Play 47
Bell's Shells 48
An Even Odder Game 49
Figuring Out the Trick 50
Wave of Chance 52
The Clone Armies 54
The Undocumented Feature 56

5 The Standard Model of Particles **59**

Zen and the Art of Particle Physics 60
A Tale of Two Particles 61
Particles of Matter *62*
Particles of Force *63*
Particles and Fields 65
Virtual Reality 66
The Weird Nuclear Force 67

6 The World of the Small **71**

Small Is Different 71
Getting a Grip on Particles *73*
Particles Come Out *75*
Particle Groupies 76
Two Types of Forces 80
The Basement of Reality 81
The Hierarchy of Nature 82

Part 3: The Need for Unity **85**

7 Why Unify? **87**

Theories of the World, Unite! 87
Woe with Einstein 89
That Empty Feeling *89*
Trying to Find the Time *91*
The Standard Model Gets Ratty 92
Hierarchy Problems *93*
A Matter of Antimatter *96*
A Punch in the GUT 96

8 Black Holes **99**

Down the Drain 99
How to Make a Black Hole—or Not *101*
Types of Black Holes *101*
Black Hole Geography 102
Taking the Plunge *102*
A Singular Problem *105*
Trouble on the Horizon *106*

The Quantum Trap Door 106
The Hawking Effect *107*
Hints of Quantum Gravity *110*

9 The Big Bang 111

Roots 111
The Meaning of the Bang *113*
Cosmic Expansion *114*
Unwinding the Clock 116
Cosmic Inflation 119
The Ultimate Beginning 121
The Dark Side 122

10 Time Machines 125

Blueprint for a Time Machine 125
Wormholes *126*
Negative Energy *129*
What's Wrong with Time Machines? 130
The Role of Quantum Gravity 132

Part 4: Gravity Meets the Quantum 135

11 The Paradox of the Graviton 137

The Primacy of Quantum Theory 137
Meet the Graviton 138
Putting a Spin on It 140
Caught in an Infinite Loop 141
There's Too Much Room at the Bottom 143
When the Ground Comes Alive 144

12 The Music of Strings 147

To Do Is to Be 147
A Tangled Tale 148
What Are These Strings, Anyway? 150
The Inner Life of Strings 152
Gravitating to Strings 153
Brane Bogglers 155

13 Playing a Different Tune 159

What Else Is out There 159
Loop Quantum Gravity 160
Loop-d-Loop *160*
Atoms of Space *161*
Where Do Loops Stand? *163*
"Buckyspace" 163
Domino Theory 164
A Tipping Point? 166

Part 5: The Big Ideas 169

14 Extra Dimensions 171

Headed in a New Direction 171
From Flatland to Hyperspace 173
Running Out of Space 175
Escaping the Shackles 177
Dimensional Shadow Puppets *178*
Footloose Gravity *178*
On the Funky Side 180
Goldilocks and the Three Dimensions 183

15 Parallel Universes 187

So Many Ways to Make a Universe 188
Planning for Every Contingency 189
Making the Possible Real 191
Level 1: Space Beyond Our Horizon *192*
Level 2: Bubble Universes *193*
Level 3: Quantum Many Worlds *194*
Level 4: The Mathematical Universe *195*
The Best of All Possible Worlds 196

16 The Root of the Tree 199

Swallowing Its Tail 199
Living Off the Grid 202
Dualing Points of View *202*
Is Anything Smaller Than Strings? *205*
Loops, Trees, and Sprinkles 206
Does Relativity Fail? 208

17 Symmetry **211**

Beauty Is Deep 211
Types of Symmetries 213
Who's the Most Symmetrical of Them All? 218
Super Well Hidden *219*
Pros and Cons *220*
A Higher Point of View 222

18 Emergence **225**

Emerging Ideas 225
Reach Out and Touch Someone 227
The Holographic Principle 228
Down the Memory Hole *229*
Adventures on the Holodeck *230*
Stringy Holography *231*
Seeing Spooky Action? 233

Part 6: What Has String Theory Done for You Lately? **237**

19 Black Branes and Balls of String **239**

Getting Warm 240
Melting Pot 241
A Black Hole Built of Branes 243
Loop Hole 245
Timed Out? 246

20 Before the Big Bang **249**

Time Before Time 249
Living With Inflation 251
Stringy Inflation *251*
String Gas and Black Hole Fluid *252*
The Cosmic Inflection Point *253*
Follow the Bouncing Brane *254*
Loopy Cosmology 256
Creation *Ex Nihilo* 257
Darkness Falls 258

21 Ten Ways to Test String Theory 261

Testing Times ... 262
What Is Proof? ... 264
1. The Large Hadron Collider ... 265
How the Collider Works ... 265
What It Looks For ... 266
2. Testing Dr. Einstein ... 268
3. Catching Some (Cosmic) Rays ... 269
4. Written on the Sky ... 270
5. Gravitational Wave Detectors ... 271
6. Watching Protons Fall Apart ... 272
7. Seeing Dark Matter ... 272
8. Cosmic Strings ... 273
9. Tabletop Gravity ... 274
10. Hints of Other Universes ... 274

22 The String Wars 277

String Theory and Its Discontents ... 277
The 20 Years' Wars ... 278
Making Sense of the Complaints ... 280
"It's Taking Too Long" ... 280
"It Can't Be Tested" ... 281
"It Can't Explain Anything" ... 282
"It Presumes the Shape of Spacetime" ... 282
"It Suffers from Groupthink" ... 283
"My Theory Is Better" ... 283

23 What Now? 287

Something's Missing ... 287
What Is Time? ... 289
Why the Quantum? ... 290
For Philosophy ... 293
The Comprehensibility of the Cosmos ... 294

According to superstring theory, what we think of as a point in our ordinary space may actually be a complex origami in six further dimensions, so tightly wrapped that it's very hard to detect the complex geometry.

This theory is not yet "battle tested" by experiment. But many are betting on it almost for aesthetic reasons. According to Ed Witten, the acknowledged intellectual leader of the subject, "Good wrong ideas are extremely scarce, and good wrong ideas that even remotely rival the majesty of string theory have never been seen."

One thing, however, is sure: string theory is hugely complicated, and is a challenge to the world's best mathematicians. Indeed, human brains could be intrinsically inadequate, and may never bring the theory to completion. But what about nonhuman minds? If we ever established contact with intelligent aliens, there might be a vast IQ gap, but there would not be an unbridgeable "culture gap." One common culture (in addition to mathematics) would be physics and astronomy. The aliens may live on planet Zog and have seven tentacles, but they would be made of similar atoms to us and would gaze out (if they had eyes) on the same cosmos.

For aliens, the intricacies of string theory may be a doddle. But for most of us humans, they are a Himalayan challenge. That is why we should welcome a book such as this—written by an expert communicator—which aims to distill the essence of these daunting ideas into a palatable brew that we can all savour.

—**Martin Rees** is Professor of Cosmology and Astrophysics and Master of Trinity College at the University of Cambridge. He holds the honorary title of Astronomer Royal and also Visiting Professor at Imperial College London and at Leicester University.

Foreword

Ever since the classical Greek era when Earth, Air, Fire and Water were believed to be the substance of the world, scientists have sought a "unified" picture of all the basic forces and "building blocks" of nature. They have sought to answer the question, "What are we, and the world, made of?" And ever since Isaac Newton showed that the force that makes an apple fall is the same force that holds planets in their orbits, scientists have tried to "unify" previously unconnected concepts.

During the twentieth century, physicists developed quantum theory. We have thereby come to understand that the essence of all substances—their colour, texture, hardness, and so forth—is set by their structure, on scales far smaller than even a microscope can see. And the work of Einstein and his successors has deepened our understanding of gravity—the force that governs the motions of planets, stars and galaxies.

But there is important unfinished business. The quantum theory, which governs the very small, and Einstein's theory, which governs the very large, haven't been meshed together into a single unified story.

In most contexts, the lack of a unified theory does not impede us because the domains of gravity and of the quantum do not overlap. Astronomers can ignore quantum fuzziness when calculating the motions of planets and stars. Conversely, chemists can safely ignore the gravitational force between individual atoms because it is nearly 40 powers of 10 feebler than the electrical forces. But at the very beginning of our cosmos—in a "big bang" when everything was squeezed smaller than a single atom—quantum fluctuations could shake the entire universe. To confront the overwhelming mystery of what banged and why it banged, we need a unified theory of the cosmos and microworld. Indeed, without such a theory, we won't understand the real "atomic" nature of space itself.

Einstein himself worked on his unified theory until his dying day. In retrospect, his efforts were doomed because little was then known about the forces that hold atomic nuclei together, and because he was famously dissatisfied with quantum mechanics. He lived until 1955, but cynics have said that he might as well have gone fishing from 1920 onward. But there's something rather noble about the way he persevered and "raised his game"—reaching beyond his grasp.

But the quest for a unified theory is no longer premature: it now engages young scientists—not just grand, ageing figures who can afford to risk overreaching themselves.

Just as all material has an atomic structure, theorists believe that space and time are themselves structured on some tiny scale—a trillion trillion times smaller than atoms.

Introduction

When it comes to string theory, the leading proposal for an explanation of the deepest workings of nature, even the world's experts feel like idiots. That's what makes it so exciting! Nathan Seiberg, a pioneering string theorist, once told me: "The theory is constantly more clever than the people who study it." Physicists are still working through what string theory means and what it implies. Starting from a few ideas and mathematical equations, the theory opens up a universe of possibility that never ceases to surprise physicists. The same is true of other contenders for a fundamental theory. We are lucky to live in such an intellectually fertile time.

One thing most people don't realize about physicists is how many of them are outdoors people, as often found on a hiking trail or a sailboat as in front of a blackboard. And that's not unconnected to their work. Physics, like all science, flows from an appreciation of the beautiful complexity and intricacy of the natural world. Physicists' role in the grand scheme of science is to figure out how this complexity can flow out of a few basic principles, as it appears to do. Their goal is to explain the universe simply yet comprehensively. This book has the same goal. I'm a firm believer that the principles of science can be boiled down to their essence without boiling them away altogether.

Although the book emphasizes string theory because it's the approach taken by the majority of today's theorists, the theory is still very tentative and could well turn out to be flat-out wrong. I try to be fairly evenhanded about the range of proposed theories. Proponents of each theory will probably complain that I've treated the others too glowingly, but I think each approach deserves a fair hearing. The science is unsettled; all the approaches have pros and cons, and none has been subjected to a real experimental test. It's a fair bet that none of them is quite right and that the ultimate theory will combine insights from all.

How This Book Is Organized

Part 1, "String Theory in a Nutshell," is a book-within-the-book. At the risk of giving away some of the punch lines, it summarizes the ideas that the rest of the book will flesh out. Not least, it tries to articulate why string theory and the rest of fundamental physics matter to begin with, something that I don't think physicists often do, even to themselves.

Part 2, "The Great Clash of Worldviews," brings you up to speed on current theories: Albert Einstein's theories of relativity and quantum theory. Right now, those theories encapsulate everything that physicists know about the world. Einstein's masterpiece, his general theory of relativity, is the archetype of a beautiful theory to which all physicists aspire. It explains so much with so little. Quantum theory culminates in the Standard Model of particles, which underlies all things chemical and biological. The model predicts that the forces of nature act differently on microscopic scales, a trend that initially seems to make nature more complicated but in the end points the way to a radical simplification.

Part 3, "The Need for Unity," describes where these current theories go astray. Relativity theory predicts that matter reaches an infinite density inside black holes and at the start of the big bang—and then leaves physicists in the lurch, unable to say what happens then. Quantum theory is so shot-through with riddles that physicists regularly hold conferences just to classify them all. The theories' faults tend to be complementary, suggesting the answers lie in their union.

Part 4, "Gravity Meets the Quantum," discusses what a unified theory might be, focusing on the question of how to explain gravity in terms of a so-called quantum theory of gravity. String theory is the foremost approach but not the only one. I explain why such a theory has been so hard to construct and how each of the proposed theories attempts to overcome these difficulties.

Part 5, "The Big Ideas," is the heart of the book. It offers some thoughts on what might go into a more complete worldview, drawing out the core ideas and the contributions of all the proposed theories. One common theme is a huge expansion in our conception of reality. Although the theories themselves are paragons of simplicity, their implications aren't.

Part 6, "What Has String Theory Done for You Lately?" goes into some of the practical importance of the theories for understanding black holes, the big bang, and the mysteries of quantum theory, as well as how these theories might have observable consequences. I also discuss some of the criticism that string theory, in particular, has come in for.

A few themes crop up throughout the book like prairie dogs rearing their heads every so often. I think that even folks who have read other books on physics will find them provocative:

- **Everything that happens, happens for a reason.** This is known as the concept of determinism. Current theories of physics flirt with randomness and unpredictability, but they are deterministic at heart. Physicists may ultimately find that things happen for no reason at all, but so far, every hint of that has turned out to be a sign of incomplete theoretical understanding.

- **The distinction between cause and effect is fundamental.** Did you yell at your brother because he took your peanut-butter sandwich, or did he take your sandwich because you yelled at him? In most households, these disputes seem unresolvable, but physics says there's always an absolute right answer. Whenever physicists construct a theory predicting that some people see event "A" cause event "B," while other people see event "B" cause event "A," the theory turns out to be inconsistent.

- **Space and time are composed of some deeper building blocks.** In many situations, far-flung particles can remain connected as if the space separating them didn't matter. This may be a glimpse of an underlying reality that goes beyond space.

For me science is like Tabasco sauce, Cuban timba music, a coral reef, or my daughter's giggle: one of the great pleasures in life, adding texture to my experience on this planet. Knowing a bit of how the world works makes it that much easier to bear its idiocies and injustices. If nothing else, I hope to convey the enthusiasm I feel for this area of science and encourage you to explore it on your own.

Sidebar Descriptions

Here are explanations of the sidebars that have been provided to offer you readily available explanations of key ideas, concepts, thoughts, and terms in string theory.

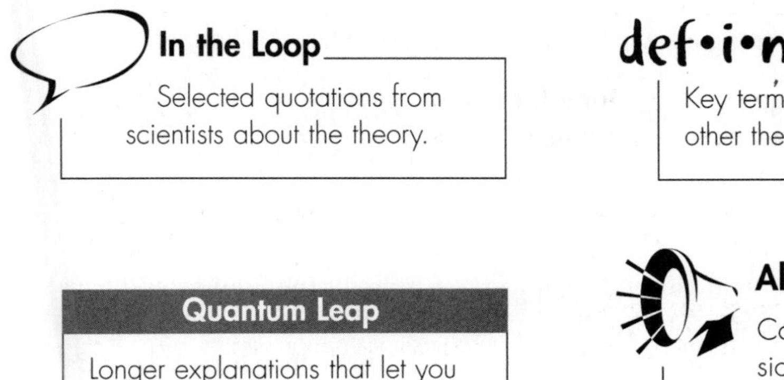

Acknowledgments

The great privilege of writing about physics is that physicists are so generous with their time and ideas. I owe a special debt of gratitude to Keith Dienes, who not only helped me think through some of the most difficult chapters but also offered to serve as the volume's technical editor, reading the whole manuscript to save me from my own idiocy. I'm also indebted to Carlo Rovelli, who guided me not only through quantum theory and loop quantum gravity but also through the city of Marseille and his own seaside village of Cassis. Others who went above and beyond the call of duty include Raphael Bousso, Sean Carroll, Joe Lykken, Nati Seiberg, and Max Tegmark. The book wouldn't have been possible without them.

I owe many people a fancy dinner for the time and care they took to read passages for sins of commission and of omission: Giovanni Amelino-Camelia, Tom Banks, Massimo Blasone, Martin Bojowald, Robert Brandenberger, Cliff Burgess, Robert Caldwell, Craig Callender, Juan Collar, Tamara Davis, Ted Erler, Glennys Farrar, Maurizio Gasperini, Shelly Goldstein, Ted Jacobson, Chang Kee Jung, Marc Kamionkowski, Gordy Kane, Paul Kwiat, Bob Laughlin, Avi Loeb, Bob Loeb, Renate Loll, Samir Mathur, Travis Norsen, Renaud Parentani, Joe Polchinski, Helen Quinn, Moshe Rozali, Bernard Schutz, Steve Shenker, Rafael Sorkin, Paul Steinhardt, Bill Unruh, Vlatko Vedral, Edward Witten, and Wojciech Zurek. If I've still erred, it's not for want of their help.

In thinking through the subject, I've benefited hugely from conversations with Scott Aaronson, David Albert, Nima Arkani-Hamed, Çaslav Bruckner, John Donoghue, Michael Duff, Larry Ford, Brian Greene, David Kaiser, Bernard Kay, Michelangelo Mangano, Nick Mavromatos, Eric Mayes, Fernando Quevedo, Stuart Raby, Richard Taylor, and Gabriele Veneziano.

Scientific American's managing editor, Ricki Rusting, showed a superhuman level of understanding when I turned up for work bleary-eyed after a long night of book-writing. I'd also like to thank Wolfram Research for *Mathematica*, the software package that makes mathematics even more of a joy than it already is. I used it to generate many of the figures.

Without the consistent support of the acquisitions editor at Alpha Books, Tom Stevens, and the development editor, Susan Zingraf, I doubt the book would have gone anywhere. And it definitely wouldn't have gone anywhere were it not for a person who read every single word, helped me tune up the language, and kept up more confidence in me than I had in myself. Marrying Talia was the least idiotic thing I ever did.

Special Thanks to the Technical Reviewer

The Complete Idiot's Guide to String Theory was reviewed by an expert who double-checked the accuracy of what you'll learn here, to help us ensure that this book gives you everything you need to know about string theory. Special thanks are extended to Keith Dienes.

Trademarks

All terms mentioned in this book that are known to be or are suspected of being trademarks or service marks have been appropriately capitalized. Alpha Books and Penguin Group (USA) Inc. cannot attest to the accuracy of this information. Use of a term in this book should not be regarded as affecting the validity of any trademark or service mark.

String Theory in a Nutshell

What makes string theory and other cutting-edge theories so compelling that physicists would turn down well-paid jobs on Wall Street to work on them? The beauty of these theories, as with science in general, is that they take us beyond ourselves. They reveal a richness to nature that we might never have suspected. String theory and its alternatives wriggle deeper in the workings of the material world than any previous theory—so deep, in fact, that there may be nothing deeper.

Chapter 1

What Is String Theory?

In This Chapter

- Introducing string theory
- Introducing loop quantum gravity
- Small means big
- Where the theories stand

Studying physics is sort of like eating an artichoke. You pull off the layers of reality and slowly get to the heart of it all. Over the centuries, physicists have been able to explain an ever-wider range of phenomena with ever-fewer laws, and they now seem to be zeroing in on the root essence of the natural world. String theory and alternative theories are the latest steps in this effort—and maybe its culmination. This book aims to explain why these theories are potentially so revolutionary, what eye-popping things they reveal, and what problems remain to be cracked.

The Ultimate Symphony

One of the joys of childhood for youngsters is to play "Stump Your Teacher." It's a game students can always win and wise teachers encourage.

When the teacher says that everything is made up of atoms, the bright student asks, "So what are atoms made of?" When the teacher replies that they're made of subatomic particles called protons, neutrons, and electrons, the student asks, "What are protons and neutrons made of?" As the teacher answers "Even tinier particles called quarks," the student then wants to know, "What are quarks and electrons made of?" At that point, the student wins. Not even the greatest expert in the world can answer that question. It's the frontier of human knowledge.

String theory lets teachers win one more round of the game. It proposes that subatomic particles are sub-sub-subatomic strings. If we zoom in on the particles closely enough, what we usually think of as little billiard balls reveal themselves to be tiny loops or lengths of a more primitive material. These strings vibrate like miniature guitar strings, and each type of particle corresponds to a string playing a certain pitch—as though quarks were middle C, electrons were E flat, and the world around us were a symphony of unimaginable intricacy.

def•i•ni•tion

String theory proposes that matter, force, space, and time are composed of tiny vibrating strings. It's widely considered the leading candidate for a unified theory of physics, which boils down all the forces and types of matter to a single set of principles.

Quantum theory describes the behavior of objects based on the assumption that matter and force come in indivisible units.

String theory unites not only the types of particles, but also the ways they behave. Currently, physicists must make do with an uneasy "shotgun marriage" of two explanations for the behavior of matter. Most phenomena, such as electricity and magnetism, fit into the conceptual framework known as *quantum theory*. But gravity stubbornly refuses to go along. It falls under the rubric of Albert Einstein's general theory of relativity.

Quantum Leap

This book uses scientific notation for very small and large numbers for the simple reason that you'd go blind if I kept writing 10^{-34} meter as 0.0000000000000000000000000000000001 meter. Besides sweeping away all those zeroes, scientific notation makes it easy to see the ratios between numbers. In physics, ratios are usually what matter. As much happens in the interval between 10 and 100 as in the one between 100 and 1,000—both involve a ratio of 10 to 1. In scientific notation, these numbers are 10^1, 10^2, and 10^3. The exponent increases in even steps, representing the equivalence of the intervals.

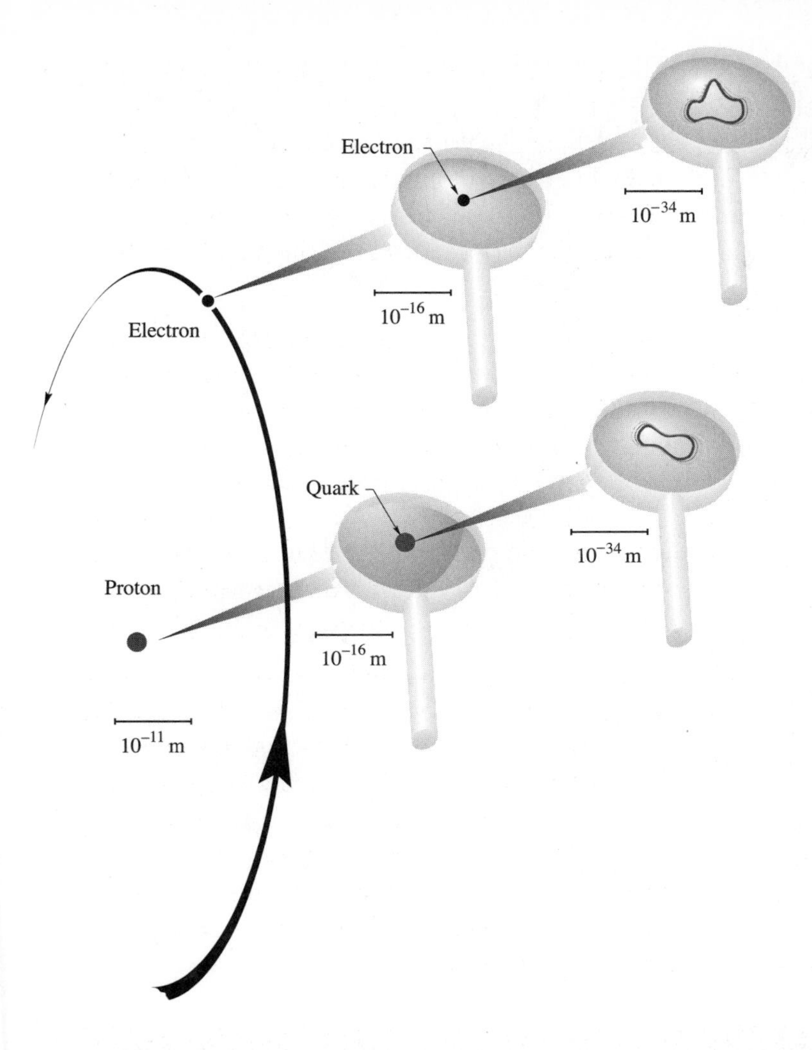

The hydrogen atom consists of an electron and a proton, each of which, according to string theory, ultimately consists of vibrating strings.

(Courtesy of Terry Anderson and Lance Dixon, Stanford Linear Accelerator Center)

The reason for this split is that gravity is special. Whenever an object exerts a force on another object, the force travels through the space between those objects. But gravity does more. It also warps space. Gravity is like a truck that doesn't just drive down a road but also causes the road surface to buckle as it does so.

To bring gravity into the quantum framework requires a theory that can handle this special feature, a *quantum theory of gravity*. Such a theory converts the shotgun marriage into a true union. Because of the

def•i•ni•tion

A **quantum theory of gravity** describes the force of gravity using quantum principles, thereby uniting standard quantum theory with Albert Einstein's general theory of relativity.

connection between gravity and the shape of space, a quantum theory of gravity would also be a quantum theory of space. Space might be far more complex than we give it credit for, like a road that looks smooth and unbroken from a distance but cracked and gnarled when viewed up close.

Space looks smooth to us but could have a complex shape on fine scales.

(Copyright 1991 Sigma Xi. From American Scientist *magazine, November/December 1991)*

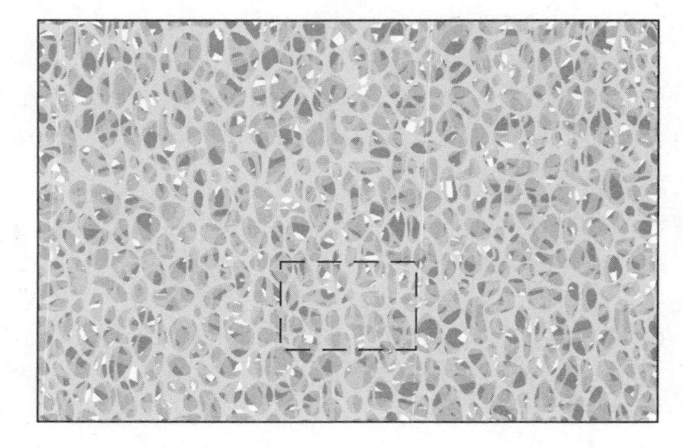

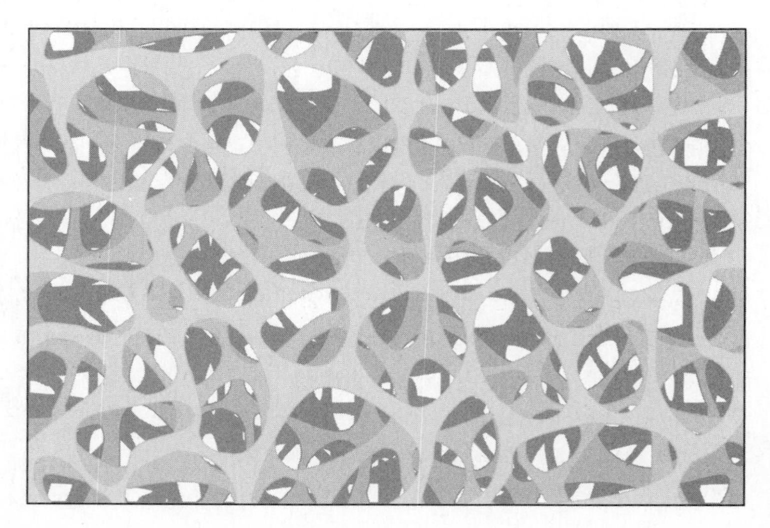

String theory fits the bill. It explains the workings of gravity as one of the ways strings vibrate. In string theory, space and matter are inseparable. Matter would be nowhere without space. Less obvious, we can't have space without at least the possibility of matter or else gravity wouldn't behave consistently.

In the Loop

The very *raison d'être* of matter is its indispensability for curing the … inconsistencies of quantum gravity.

—Hermann Nicolai, Kasper Peeters, and Marija Zamaklar, Max Planck Institute for Gravitational Physics

Alternative Music

String theory gets the most attention but is by no means the only contender for a deep theory of nature. Whereas string theorists see gravity as the lost sheep and seek to bring it back into the particle flock, physicists who prefer the leading alternative, known as *loop quantum gravity*, see gravity as the sheep dog. To them, the special features of gravity demand special treatment.

def•i•ni•tion

Loop quantum gravity (or simply "loop gravity") is the leading alternative to string theory for a quantum theory of gravity. It describes space in terms of linked atoms of volume.

According to this theory, space consists of atoms—not ordinary atoms, but little chunks of space that can't be subdivided into anything smaller. Although loop gravity doesn't set out to explain all the particles of nature, some of its proponents think it might still explain ordinary particles as little bits of tangled space, like knots in a carpet. Astronomer Carl Sagan famously said that we are all made of starstuff—chemical elements created deep within stars. Both string theory and loop gravity suggest we are made of spacestuff.

The curious student might press on and ask, "So what are strings or atoms of space made of?" Physicists can't answer that yet. These things might turn out to be the truly fundamental building blocks of the world, or they might be an approximate way of describing a still-deeper level of reality. The student who wants to know will have to join the effort to find out.

Big Things Come in Small Packages

Both the strings of string theory and the space atoms of loop gravity are small—incredibly small. By most estimates, an atom is to one of them what the entire observable universe is to a human being. No conceivable microscope will ever take a picture of a wriggling string. But physicists don't really care about the building blocks, per se. They're after the principles that govern our world, and they zoom in on the microscopic level simply because that's where the principles are laid bare.

In this, physics is like any domain of life where the guiding principles seem distant to us. What does it matter that we live in a democracy, for example? We don't vote very often, and even when we do, our individual participation hardly affects the outcome. The principles of democracy don't put food on your table or play rhythm guitar in your band. But without them, you might not even have a table or a band.

The same goes for the principles of physics. They set the basic parameters of our existence, starting with the fact that we exist at all. When scientists centuries ago conjectured that the world is made of atoms, many people thought those tiny scraps of matter were abstractions that are irrelevant to our lives. The technology of the day could never hope to observe them. Yet the nature of atoms is essential to everything we see and do. If the world weren't built of atoms, chemical reactions would fail to operate and life would be impossible.

Likewise, strings or something else that fulfills a similar role are essential to understanding how the world is put together. Without them, space and time might not even exist. Objects would have no location and events would have no duration; our world would be a static, structureless mush.

So although strings may be small, the principles they embody are anything but. Exactly what those principles are, physicists aren't yet sure. What they do know is that the principles are going to be revolutionary. The reason is that unifying quantum theory and general relativity isn't simply a matter of force-fitting a few equations together. It requires two profoundly different ways of looking at the world to be reconciled, a task that famously stymied Einstein himself and has challenged every physicist since his time. Each of these worldviews has its failings, but each also has an integrity to it. It's not at all obvious how to fix their faults without wrecking their successes. It takes some new conceptual input, some novel idea that human beings never before realized or appreciated.

In the Loop

The beginning of the twenty-first century is a watershed in modern science, a time that will forever change our understanding of the universe. Something is happening which is far more than the discovery of new facts or new equations. This is one of those rare moments when our entire outlook, our framework for thinking, and the whole epistemology of physics and cosmology are suddenly undergoing real upheaval.

—Leonard Susskind, Stanford University

The different approaches to unification have varying degrees of ambition, but in some way or another, the unified theory will cover every phenomenon known to physics. Because of its scope, the theory will go right to the foundations of physical reality, and it will probably be unlike anything science has ever seen. Physicists get around the limitations of both relativity and quantum theory by saying that some deeper theory will explain them. A fully unified theory won't be able to pass the buck.

The conditions required to make such a theory work are so stringent that only a single set of concepts might be able to satisfy them. There might be no other way for a universe to hang together.

String theory comes closer to achieving this goal than any other effort that physicists have ever made. It's not there yet, and it may well turn out to be completely wrong, but what encourages string theorists is that if we work through what it takes for a string to vibrate, it can do so only under very specific conditions. An ordinary guitar string doesn't encounter the same restrictions. It's so large and floppy (by physics standards) that the counterintuitive aspects of relativity and quantum theory don't come into play. For a miniature string, though, things get more complicated—which is good, because the restrictions on its behavior serve as an organizing principle of nature. In this way, string theory helps us make sense of a world that so often seems senseless.

String Instruments

New principles always reveal themselves grudgingly. Consider how Einstein's theories of relativity came about. The nineteenth-century experiment that paved the way for his theories—by discovering that light moves at a constant speed, independent of the speed of whatever emits it—had a precision of about 1 part in 10,000. Later, Einstein's ideas about gravity were borne out by the shift in a position of a star on a photograph by little more than a hundredth of a millimeter.

In fact, there's a sort of inverse relationship: the broader the conceptual revolution, the harder you have to hunt for it. After all, if the new principles were so obvious, people would have noticed them already.

For string theory, loop gravity, or whatever other explanation emerges for the innermost workings of nature, the predicament is acute. Relativity and quantum theory make predictions that agree with observations, some as precise as 11 decimal places. This empirical success makes finding a new theory all the more difficult. The answer may lie in strings, but strings are small and their direct effects are proportionately tiny.

Just managing to combine relativity and quantum theory into a single theory is a step. Any theory that unites them inherits their observational successes. But physicists also seek distinctive predictions—ways that strings go beyond what we already know to reveal unanticipated aspects of the universe, something about the world that people had been blind to.

It could be a deviation in the twelfth decimal place, or even further down the line. Not only must instruments have the precision to test such a prediction, they must have the discriminating power to distinguish it from a thousand confounding effects with no deeper significance.

Or there could be some other subtle clue that was staring us in the face all along. In ordinary life, it pays to be attuned to subtle clues. The "broken windows" theory in sociology is an example. A broken window that hasn't been fixed or graffiti that hasn't been scrubbed away seems fairly minor on its own. But the fact that people don't attend to these little things hints at deeper problems. The poet William Blake gave the canine version:

> A dog starv'd at his master's gate
>
> Predicts the ruin of the state.

For someone to let his pet go hungry, something must be very wrong in a society—maybe economic hardship or a cycle of violence. To the trained eye, a seemingly minor occurrence is a sign of a much broader question of principle.

Many physicists have worried whether such clues even exist for string theory and the other proposed theories. But things have been looking up lately. A number of new scientific instruments are able to test aspects of unified theories. The best known is the Large Hadron Collider, the largest particle smasher ever built—in fact, the largest and priciest scientific instrument of any kind. The collider is looking for novel phenomena and, if string theorists' most optimistic predictions are right, could create fleeting black holes and peer into higher dimensions of space. New space-based astronomical observatories, such as NASA's GLAST satellite and the European Space Agency's Planck satellite, are starting to check what space looks like on its very finest scales.

The giant ATLAS particle detector at the Large Hadron Collider.

(Copyright 2005 CERN)

Any discoveries by these instruments will involve incredibly tiny effects: one particle in a billion that acts up or two particles that race neck-and-neck for billions of years across the universe only to arrive a millisecond apart. And even those teensy signals don't get at the core of string theory or loop gravity. No feasible instrument has anywhere near the resolution we'd need to prove or disprove either theory for sure. But proof or disproof in science is seldom so clear-cut. A theory steadily accumulates points in its favor or points against, until physicists judge that their time is better spent on something else. The next few years could prove decisive in either solidifying string theory or knocking it out.

The fact that string theory and other such theories are works in progress is what makes them so exciting. We are watching ideas come together before our very eyes.

Every generation thinks it lives in a special time, but a quantum theory of gravity could be *our* era's claim to specialness. If one of the proposed theories works out, it will be one of the things future generations remember about us.

The Least You Need to Know

- String theory is the idea that subatomic particles are little vibrating strings.
- String theory also seeks to explain space and time.
- This is cutting-edge science, not a done deal, and in fact, string theory may well prove to be wrong.
- Another proposed theory to explain the deep workings of nature goes by the name of loop quantum gravity.
- Both string theory and loop quantum theory predict a wealth of new phenomena, implying that humans have barely scratched the surface of what's out there.

Chapter 2

Why Should Anybody Care?

In This Chapter

- The meaning of theory unification
- The obvious that's not so obvious
- Being open to a conceptual revolution
- The coexistence of science and religion

The way some people describe string theory and similar efforts—as an exercise in abstract mathematics, postulating entities as small in relation to an atom as we are to the known universe—you could be forgiven for wondering why anyone should care. Even theorists themselves have been guilty of downplaying their own theory. However, although the strings or loops may be small, they imply something huge: a radically new view of the world.

The Tree of Physics

Physics deals with some of the deepest questions one can ask, like: Why does toast always fall with the buttered side down? How can you avoid pocketing a cue ball when playing pool? Why is my desk always such a mess?

You don't need to jump straight to big questions such as the nature of space and time. These little mysteries of everyday life lead you to them, since every "because" leads to another "why." The reasons (the because) could involve Isaac Newton's laws of motion and gravity or the properties of materials. Then you ask why those laws hold, and the laws beneath them, and so on. As you descend further and further, a remarkable fact about our universe emerges: phenomena that seem completely different have related explanations.

People used to think two distinct principles governed the fall of an apple and the cycles of the moon, but Newton traced both to the same basic law of motion. And people used to regard magnetic attraction, static electric shocks, and light beams as unrelated, but the nineteenth-century physicist James Clerk Maxwell showed they are aspects of one single force called electromagnetism. Nature is like a big, bushy oak tree. If you start at one of the leaves—the observable phenomenon—and follow the twig to the branch to the limb, you'll discover they link together.

Nature is rich and complex; look beneath its canopy and a structure becomes apparent.

(Courtesy of Jay Ryan)

No matter which leaf you start with, you'll eventually wind up on one of two main boughs: Albert Einstein's general theory of relativity, which deals with gravity; and quantum theory, which accounts for chemical and nuclear reactions. The two not only handle different types of questions but also handle them in different ways. Relativity gives answers in terms of the behavior of space and time; quantum theory speaks of the interactions of subatomic particles. By the time you reach one of these two theories, you have transformed the original mundane question into a deep one, which can be defined as one that a five-year-old would ask and the greatest expert in the world couldn't answer (never mind the parents). What are space and time? What are particles, and why do they interact as they do?

The physicists now working to develop string theory and other such theories used to be the five-year-olds who never got a satisfying answer to these questions and have been bothered about it ever since. Their ambition is to find the link between relativity and quantum theory and reach the trunk of the tree, the place where all possible questions about the physical world converge.

Take, for example, the falling toast question. As the toast slides off the kitchen counter, its weight becomes imbalanced, so it does a somersault as it falls—a consequence, ultimately, of general relativity. But at the rate it spins, it can only complete half a turn by the time it hits the floor because of the countertop height that human anatomy entails—a consequence, ultimately, of quantum theory. And here the chain of reasoning stops. Simply put, the universe is built in such a way as to ensure that toast dropped from the typical table or counter falls face down, but no one knows why. So forget about exotica such as black holes and the big bang; explaining the mess on your kitchen floor is reason enough to unite general relativity and quantum theory.

The Joy of Unification

If you didn't know much about trees, you might try to make one by taking two sticks and wrapping a rubber band around them. Ta-da: you have a tree. That's like trying to make a unified theory of physics by taking the various observations you've made and drawing one big box around them. Ta-da: you have a unified theory.

But somehow that doesn't seem very satisfying. If you describe nature as disjointed phenomena, you can't claim to know how it or each of its components ticks. The tree of physics is more than a bunch of unrelated branches because it shows the system of how one aspect of nature follows organically from another. The thing has a life of its own, and surprises await the patient explorer.

There's no better place to see the underlying order of nature than on a pool table. The three main principles of motion, the *conservation of energy*, *momentum*, and—if the ball is spinning—*angular momentum*, come into sharp focus as you hit the ball.

def•i•ni•tion

Conservation of energy means that an object or collection of objects doesn't gain or lose energy over time. **Conservation of momentum** or of **angular momentum** means that a moving or spinning object continues to move or spin unless some outside force acts to stop it.

Pool balls may slow down, but that just means the energy of their motion is converted into other forms, such as heat and sound. The total energy doesn't change. Much the same holds for momentum and angular momentum. When balls collide, they share their motion or spin but do not lose it. These conservation principles, in turn, arise because the laws of physics do not vary in space and time. Here, "space" is the surface of the pool table. If it's level, the ball won't gain or lose momentum on its own, and if the felt covering the surface is smooth, the ball will retain whatever spin you give it. More abstractly, time is a level surface in that the laws of physics don't distinguish past from future.

If you put these principles together and do some math, you can deduce that a rolling cue ball, after colliding with the target ball, heads off at an angle of about 30 degrees to its original path. (For more detail, see *The Complete Idiot's Guide to Pool and Billiards* by David Alciatore.) Make sure the pocket isn't in that direction at that angle, and you'll save yourself a lot of embarrassment. If you do scratch, you can always blame it on the universe.

Those three principles allow for a huge variety of pool games. Physicists don't say that a pool game is *just* a few principles or that the world is *just* particles, any more than biologists say that the variety of species is *just* the product of evolution. To the contrary, modern scientific theories reveal the world as a process of self-creation. From the base of the tree on up, each level adds something that didn't exist at the lower levels. If that weren't the case, if the lower levels weren't any simpler, the laws of physics would need to specify each and every aspect of the world, and where would the creativity be in that? It would be like saying the species all came into existence just as they are and stayed that way, static and passive.

The idea of *emergence*, whereby complex phenomena emerge from simple laws, is often taken to be the opposite of *reductionism*, which breaks systems down into simpler pieces. It's more productive to think of reductionism and emergence as two sides of a coin. When you take something apart, you do so not just to catalog the pieces, but also to figure out how they fit together—and then to find new things they might do. So the unity of nature is not a trivial unity or undifferentiated blandness, but rather a unity with countless forms of expression.

def•i•ni•tion

Emergence is the principle that a complex system has properties its components don't. In the words of physicist Philip Anderson, "The whole becomes not only more than but very different from the sum of its parts."

Reductionism is the principle that a complex phenomenon can be broken down into smaller pieces that are easier to explain.

The ultimate ambition of physics is to push this idea to the max: to show that *everything* we know—even things that we don't normally think of as "things," such as forces, space, and time—are all aspects of the same basic stuff, be it a wriggling string or an atom of space.

Far from reducing nature to a beige mush, the fully unified theory could reveal that the world is richer than anyone ever imagined. Seldom does a theory bring together what people already knew without opening their eyes to possibilities they'd never suspected. To go back to the tree, once you trace the leaf to the branch to the tree trunk, you can explore a new, different branch. And who knows what you might come across—maybe a three-toed lizard or an abandoned tree fort. Every past unification in physics has shown that the world we know is just a small part of what's really out there. Maxwell introduced us to forms of light beyond the range of our vision. Einstein brought us black holes. Although string theorists' toehold on the tree is still shaky, they've already caught glimpses of truly mind-blowing marvels.

Great Unifications in Physics

Year	Unifier(s)	Theory	What It Unified
4th century B.C.	Aristotle	Aristotelian natural philosophy	Matter, change, motion, and cause
1686–1687	Isaac Newton	Laws of motion and gravitation	Celestial and terrestrial motion
1861	James Clerk Maxwell	Electromagnetism	Electricity, magnetism, and light
1869	Dmitri Mendeleev	Periodic table	Chemistry
1905	Albert Einstein	Special theory of relativity	Electromagnetism and laws of motion
1915	Albert Einstein	General theory of relativity	Special relativity and gravitation
1900s–1920s	Neils Bohr, Werner Heisenberg, Erwin Schrödinger, and many others	Quantum mechanics	Electromagnetism and atomic theory of matter
1920s–1940s	Paul Dirac, Richard Feynman, and many others	Quantum field theory	Special relativity and quantum mechanics
1960s–1970s	Abdus Salam, Sheldon Glashow, Steven Weinberg, and many others	Electroweak theory	Electromagnetism and weak nuclear force

Why Is This Theory Unlike All Others?

I hate to be the one to break the news, but somewhere hidden in your thoughts is an interloper, something you've taken for granted all your life, something obvious, essential, and wrong. Scientists, too, have grown up with this assumption all their lives. Part of our shared worldview has got to go, but no one knows which.

Smoking out interlopers is one of the creative acts needed for unification. As Einstein was developing his special theory of relativity, he faced a serious dilemma. Newton's laws of motion were spectacularly successful. So was Maxwell's theory of electromagnetism. But the two refused to connect. Maxwell's equations indicated that light traveled at a fixed speed, yet Newton's laws suggested there was no such thing as a fixed speed. If the light source is aboard a moving train or planet, the light should get a boost. So what gives?

And this is the essential paradox of unification. To be worthy of unification, a theory has to be successful, but if a theory is successful, what need does it have to unite with another? The predicament is like that of two perfect people who want to marry the perfect spouse. They find each other, and you can practically hear the violins in the background. But because both think of themselves as so perfect, they are unwilling to make the compromises needed to live together.

Einstein realized the trouble was something so obvious no one had questioned it: the assumption that time is absolute, passing at the same rate for everyone. As soon as he showed this interloper the door, the two theories clicked. If time slows down as the light source speeds up, then light travels at a constant rate. In this way, science is about *undiscovering* things as much as discovering them.

The same drama is now unfolding again. Both relativity and quantum theory are spectacularly successful. Physicists do not know of a single experimental exception to either; both have solid theoretical formulations. Yet they are incompatible. They barely even speak the same language. So what gives? Whatever it is, it has to be profound, or it would have been worked out by now. Physicists have sweated over a quantum theory of gravity for nearly 90 years. It has become a multigenerational project, like the building of a cathedral. Nearly all the top physicists of the past century devoted at least part of their careers to it, and Einstein himself worked on his version of a unified theory for the last third of his life. Talk about delayed gratification!

In turn, physicists expect that quantum gravity will be even more revolutionary than past unifications. It may well entail a full unification of all the phenomena known to humans, in which case it will be the first theory without fine print, such as "use only for small particles" or "don't apply at such-and-such a time." No one knows whether it will be the "final" theory—one that needs no further explanation—or not. But it will mark the end of the reductionist strategy that has proved so productive in physics: the effort to seek explanations in terms of ever-smaller things. There will simply be no such thing as "smaller." Space and time themselves might emerge from more fundamental entities that exist beyond space and time.

In the Loop

If there is a basic length scale, below which the notion of space (and time) does not make sense, we cannot derive the principles there from deeper principles at shorter distances. Therefore, once we understand how spacetime emerges, we could still look for more basic fundamental laws, but these laws will not operate at shorter distances. This follows from the simple fact that the notion of "shorter distances" will no longer make sense. This might mean the *end of standard reductionism*.

—Nathan Seiberg, Institute for Advanced Study

If the theory is, in fact, the mother of all theories, it will somehow have to contain its own justification, to show that the universe could not have been otherwise. It will need to account for all the numerical values and starting conditions that present theories attributed to deeper theories. It might reduce to pure mathematics, something that you could in principle derive from pure logic, or maybe it will require a mental framework that goes beyond mathematics.

Quantum Leap

I put quotation marks around "final" in "final theory" because, as powerful as the theory might be, it'll leave plenty of mysteries unanswered—including some of the biggies, such as the nature of consciousness. Physics tells us what the building blocks of nature are, but just because we learn how to make bricks doesn't mean we know how to build a house.

Such a theory might be the only hope for making sense of our messy desk. The tottering piles themselves are easy enough to explain: there are more ways to be disorganized than organized, so the world around us naturally degenerates into chaos unless we expend effort to keep it shipshape. The real mystery of our desk isn't that it's messy, but that it ever used to be orderly. The answer ultimately lies in the fact that the cosmos as a whole started off in a nice, clean, crisp condition. Those starting conditions are beyond the scope of current theories; they call not just for a new theory, but for a new sort of theory. To try to keep your desk clean is to fight cosmic destiny.

Big Ideas Don't Like to Be Cooped Up

If you carefully watch the positions of the planets night after night, you'll notice that many of them periodically seem to stop moving, go into reverse, stop again, and then resume their forward motion. Humans watched this happen for thousands of years before someone, most famously Nicolaus Copernicus, realized it meant the sun, rather than Earth, was at the center of the solar system. To some people, swapping two celestial bodies sounded like an exercise in pure mathematics. But its implications quickly became apparent.

For starters, this repositioning meant our planet was only one among many. It suggested that there wasn't one set of laws that applied to stars and another to those of us stuck on the ground. Instead, a single set of laws governed all the universe. The effects of the conceptual revolution rippled outward into society, feeding the Renaissance and Enlightenment eras. Although astronomy was only one part of this broader movement, and although the seeds for this intellectual flowering had been planted long before Copernicus, the celestial rethink became the prototype for questioning authority of all kinds. As science historian Thomas Kuhn wrote in *The Copernican Revolution*, "A scientist's solution of an apparently petty, highly technical problem can on occasion fundamentally alter men's attitudes toward basic problems of everyday life."

Like individuals, a whole society can grow when forced to confront new ideas. New ideas beget newer ones, and science can play a special role in jumpstarting this virtuous cycle. It looks beyond everyday experience or shows the experience in an unexpected light, so its insights are genuinely new.

Consider how far our understanding of physics and astrophysics has come within the lifetime of the oldest people today. At the start of the twentieth century, no one knew molecules or atoms existed, let alone subatomic particles. Most of the electromagnetic spectrum, ranging from radio waves to x-rays, was a laboratory curiosity. The planets of our solar system were tiny discs of light; no one had ever seen images from the surface of another world. No one had even seen our planet as a planet: a blue marble on black velvet, coated with a fragile veneer of water and air.

In the Loop

Even most people who are enthusiastic about string theory tend to underestimate how radical it will prove to be in its impact on how we understand physical law.

—Edward Witten, Institute for Advanced Study

These concepts have led to snazzy gizmos and made a lot of people rich, but just as important, they have made the world that much more interesting a place to live in. We can expect the same from the unification of physics.

Sense and Transcendence

When I was a kid, I remember playing the game of opposites. I'd tell my friends a word and ask them what its opposite was. By shouting words fast enough, I hoped to catch them saying the opposite of "dog" is "cat" or "salt" is "pepper."

When I ask people for the opposite of "science," I sometimes catch them saying "religion." In reality, the two are as opposite as dogs and cats. They might chase each other, but they wouldn't even know what to do with their supposed adversary if they caught it. If you give them plenty of pillows to nap on, they are usually content to leave each other alone. In an obliging household, they might even be found lounging around together.

I am not a religious believer myself, but I have found that some of the most intense discussions I have had about science are with strong believers. They care. The way the world is put together matters to them. They think about it; they reflect on it. So I approach the topic of science and religion with the experience that, in an obliging household, science and religion can be found lounging together.

All Tangled Up

Does science oppose religion? It's true that science usurps what used to be a major goal of religion: providing explanations for natural phenomena. In so doing, it has made room in our culture for a secular worldview. But that is not the same as saying science proves a secular worldview. Science is agnostic. By its very nature, it is incapable of saying whether a transcendental reality exists or not. Nor can it provide a comprehensive moral code. Scientists who promote atheism are speaking not as scientists but as adherents of their own belief system.

Believers and nonbelievers alike can find much to reflect on in modern physics. Many people are struck by a resemblance between the big bang and the Biblical account of Genesis. Others draw a comparison to Kabbalah or Sura 21 of the Koran. Still others see a link between quantum theory and Hindu, Buddhist, or Taoist mysticism. There is more than one way of telling the story of the universe, drawing out the aspects that are meaningful to each of us personally. One runs into trouble only if interpretation goes on in a vacuum, cut off from evolving human understanding, in which case one needs to ask these questions: "What makes me so sure I'm right? What makes me so sure that the universe conforms to my preconceptions?"

A unified theory will give us all much to chew over. What if the deepest foundations prove to be beyond reason? Physicists might be unable to come up with a final theory no matter how hard they try, or they might develop a unified theory that seems arbitrary. Few other branches of science have such potential for clarifying the boundary between physics and metaphysics, where questions of "how" give way to questions of "why."

A Shared Effort

To feel the world clicking together in your head is one of the greatest pleasures in life. It's no surprise scientists become addicted to it. As Jacob Bronowski, the British mathematician and essayist, argued, this pleasure is one that anyone can share, because when you realize what the scientist has done, you live the original eureka moment for yourself.

This collective participation in science goes far beyond individual light bulbs going off in individual heads. The search for deeper theories of nature is hard, and people need to pool their talents. And that doesn't just mean scientists. If you traced the origin of every part in, say, a particle collider, along with the parts that make up those parts, the machines it took to make them, the support the makers of those machines needed, and the money the whole thing required, you would find it took hundreds of millions of people to bring the instrument into the world. The achievement belongs to all of us.

The Least You Need to Know

- String theory and others of its sort try to tie up or unify everything we know into one related package.
- If physicists are right, things you thought were totally different are actually deeply related.
- To create a unified theory, scientists need to pinpoint the extraneous assumptions in their current understanding.
- The theory could change our thinking in ways we can't anticipate, in the way all truly novel things do.

Part 2 The Great Clash of Worldviews

String theory and other quantum theories of gravity are the latest stage in a revolution that began early in the twentieth century with Einstein's theories of relativity and quantum theory. But the word "theories" doesn't do justice to these intellectual achievements. They're not just a bunch of geeky equations; they're entire worldviews—broad-ranging ways of thinking about nature that are based on intuitive ideas yet have deeply counterintuitive implications.

Chapter 3

Einstein's Theories

In This Chapter

- Exploring the principle of relativity
- The role of the speed of light
- Uniting space and time
- All physics is local
- The world's most famous equation
- Warping space and time

If only every theory could be like Einstein's theories of relativity. Simple yet deep, they connect down-to-earth ideas to some truly out-of-this-world phenomena. The space around us, which we normally barely even think about, comes alive.

Trains of Thought

It's a good thing Albert Einstein didn't have to do my morning commute because the great man was fond of using trains in his explanations. His trains moved, rather than getting stuck in the tunnel as mine often does. He talked about synchronizing an onboard watch to a station clock, which works best when the trains run on time, as mine never does. He imagined people shining light beams down the length of train cars, which takes superhuman coordination if the train keeps lurching. Fortunately for science, Einstein's trains were Swiss.

The beauty of Einstein's theories of relativity is that, despite their spacey reputation, they're grounded in things as earthy as trains, clocks, lightning bolts, telegraph lines, and elevators. If Einstein had awakened one day and pronounced that space and time were united and black holes were floating out in the universe, people would have asked what he'd been smoking in that pipe of his. Instead, he started from simple premises and slowly built up to radical conclusions. The power of his logic pulled people along.

The theories are twofold. The first, which Einstein formulated in 1905, is the *special theory of relativity*, which describes motion in the absence of gravity. It's the source of the famous equation $E = mc^2$. The second, from 1915, is the *general theory of relativity*, which includes the effects of gravity. It's the one that has resisted the unification of physics for so long.

def•i•ni•tion

The **special theory of relativity** is Einstein's theory of motion in the absence of gravity, which holds that the speed of light is the same for all observers.

The **general theory of relativity** is Einstein's theory of gravitation, which attributes gravitational forces to the warping of space and time.

Both theories involve thinking carefully about how situations look from different points of view. Most of us have looked out a train window and not been able to tell whether our train or the other train was moving. In a very real sense, it makes no difference. Motion is relative. As long as the train moves at a steady speed without lurching, you might never even know you're on one. You can play Nerf basketball and miss as many shots as you would anywhere else. Only the motion of the ball relative to you matters. In a more sober frame of mind, you might do physics experiments. Whether you move a magnet past a coil of wire or a coil of wire past a magnet, you get the same effect: you generate electrical current. Again, only the relative motion counts.

More broadly, although each of us may see the world very differently, the laws of motion—and of all else besides—are the same for all. The great diversity of nature reflects accidents of history or the biases of our individual points of view, rather than any fundamental differences. So says the *principle of relativity*, first articulated by Galileo Galilei in the seventeenth century. Not knowing the glories of Swiss trains, he described his thinking in terms of a Venetian sailing ship.

def•i•ni•tion

The **principle of relativity**, which underlies both of the theories of relativity, holds that the laws of physics are the same for all observers.

For Einstein, the principle was self-evident. If it didn't hold, some lucky people would see the real laws of physics and the rest of us would have to make do with a cheap knockoff. Apart from offending our egalitarian instincts, this would pose a practical problem. How do you measure velocity (the direction and speed of motion) if it's not relative? When you throw a ball at 20 miles per hour, what's its absolute total velocity? Do you need to add on 700 mph for Earth's rotation around its axis, 70,000 mph for its motion around the sun, 500,000 mph for the motion of the sun through the Milky Way galaxy, and so on? If so, who could ever hope to predict the path of a ball? Science, not to mention playing outfield, would be impossible.

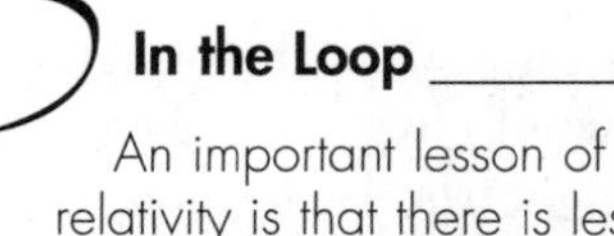

In the Loop

An important lesson of relativity is that there is less that is intrinsic in things than we once believed. Much of what we used to think was inherent in phenomena turns out to be merely a manifestation of how we choose to talk about them.

—David Mermin, Cornell University

So the principle of relativity ensures that the universe can be subdivided into manageable chunks. If it didn't hold, we couldn't understand anything without understanding everything.

All Tangled Up

Does relativity say everything's relative? Actually, what gives Einstein's theories their power is that they describe what remains the *same* despite your viewpoint: the physical laws, the speed of light, and the ordering of cause and effect. The theories are decidedly not a scientific version of moral relativism, the ideology that different value systems are equally valid.

There's No Time Like the Present

As intuitive as the principle of relativity is, it seems to lead to a contradiction. If the laws of physics are the same for everyone, then so is the speed of light. Once you set a light wave in motion, it takes on a life of its own, divorced from its source. It moves at a rate that depends only on the laws of electromagnetic waves. Experiments have confirmed this autonomy to high precision. But it throws our expectations about relative motion for a loop. Suppose you're standing on a station platform and I'm on a train gliding by you at 25 miles per hour. If I throw a ball forward at 20 mph, it'll actually be going 45 mph relative to you standing on the platform—a fact you'd better keep in mind if you try to catch it. The total velocity is the sum of the two mph's. So it stands to reason that if I shine a light beam toward the front of this moving train, it'll go 25 mph plus the speed of light. Yet you're supposed to see the light moving at the same speed as I do. So what gives?

This apparent paradox is one of those situations I discussed in the last chapter where an interloper is hidden in our thoughts, some unstated assumption we've taken for granted and shouldn't have. Einstein, standing on the shoulders of Dutch physicist Hendrik Lorentz and others, smoked it out. The total velocity is not simply the sum. A new rule for calculating relative velocities ensures that light always moves at the same speed for everybody. And that means something funky must be going on with time.

Suppose my fellow passenger and I stand back-to-back, like duelers, at the center of the moving train car. Each of us throws a ball toward the end of the car. The balls will hit the wall at the same time. You, standing on the station platform, also see the balls hit at the same time. Although the train car is moving, the balls are moving along with it.

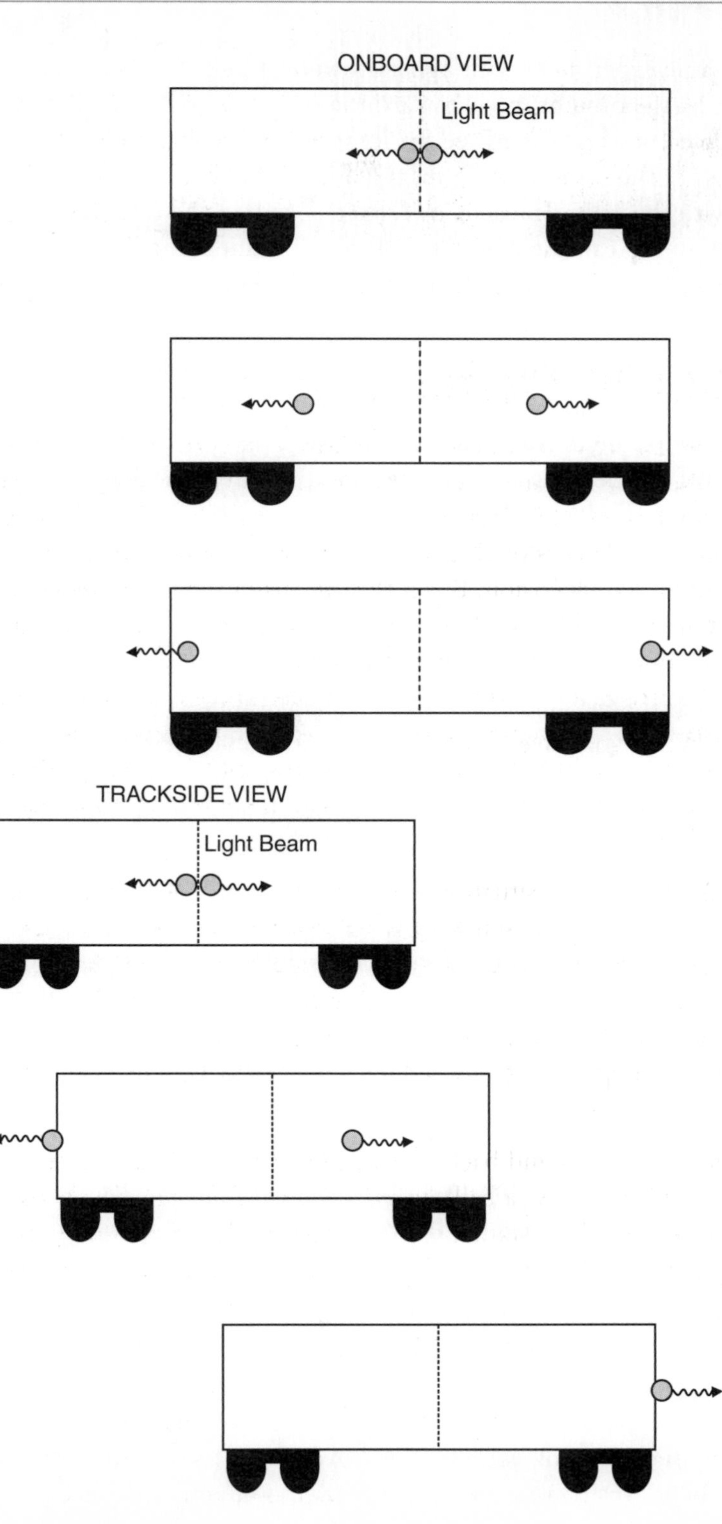

Light beams emitted from the center of a train car hit both ends simultaneously (as seen by a passenger) or the rear first and then the front (as seen by someone on the ground).

Now suppose I do the same experiment with light. I turn on a light bulb in the middle of the moving train car. Shafts of light travel toward the two ends of the car. A fraction of a second later, those of us on board see the light beams reach the ends of the car. But to you, standing on the station platform, the outcome differs. Unlike the balls, the light isn't moving with the train. It has its own fixed speed. In the time it takes light to travel the length of the car, the rear comes forward to meet the light, and the front of the car moves away from it. Accordingly, you see the light hit the rear before its hits the front.

What's simultaneous for me—light hitting the front and rear of the car—isn't for you. We disagree about the timing. Who's right? The principle of relativity says we both are. Otherwise we're left with the uncomfortable conclusion that one person is correct and the other has gone off his medicine again. One way to resolve the disagreement is to argue that the person on the ground has the truer perspective. We are all accustomed to thinking of the ground as our reference, relative to which all other motion is measured. But who's right in a perfectly symmetrical situation, such as two trains whooshing by in opposite directions or two spaceships passing in the depths of the galaxy, with nothing to serve as an external reference point?

Relativity theory holds that the resolution of our disagreement must lie elsewhere. As Sherlock Holmes said, "When you have eliminated the impossible, whatever remains, however improbable, must be the truth." In this case, the culprit must be the assumption that we all share the same notions of "now." Why should we? What reason do we have to assume that our view of the present moment extends to everyone else? All we really share is the same laws of physics. Our notion of "now" is a secondary concept based on how we probe the world using clocks and light signals. And when we probe carefully, we find that our perceptions of the present moment do not, in fact, match.

Slow Down, You Move Too Fast

As if that weren't enough, our clocks don't even run at the same rate. Suppose I'm on my train again, this time holding a yardstick vertically. I flip the overhead light on and measure how long it takes the light pulse to travel the length of the stick. This "time" is simply one yard divided by the speed of light, which comes out to about 3 billionths of a second.

For you, standing on the platform, the yardstick moves forward during those billionths of a second. For the light to travel along the stick, it, too, must move forward in addition to downward. So for you, the light travels at an angle. This additional amount of movement means that light travels a longer distance in all. Its overall speed doesn't change, and that can only mean it took extra time to cover the distance. So what my watch measures to be 3 billionths of a second, yours shows to be rather longer.

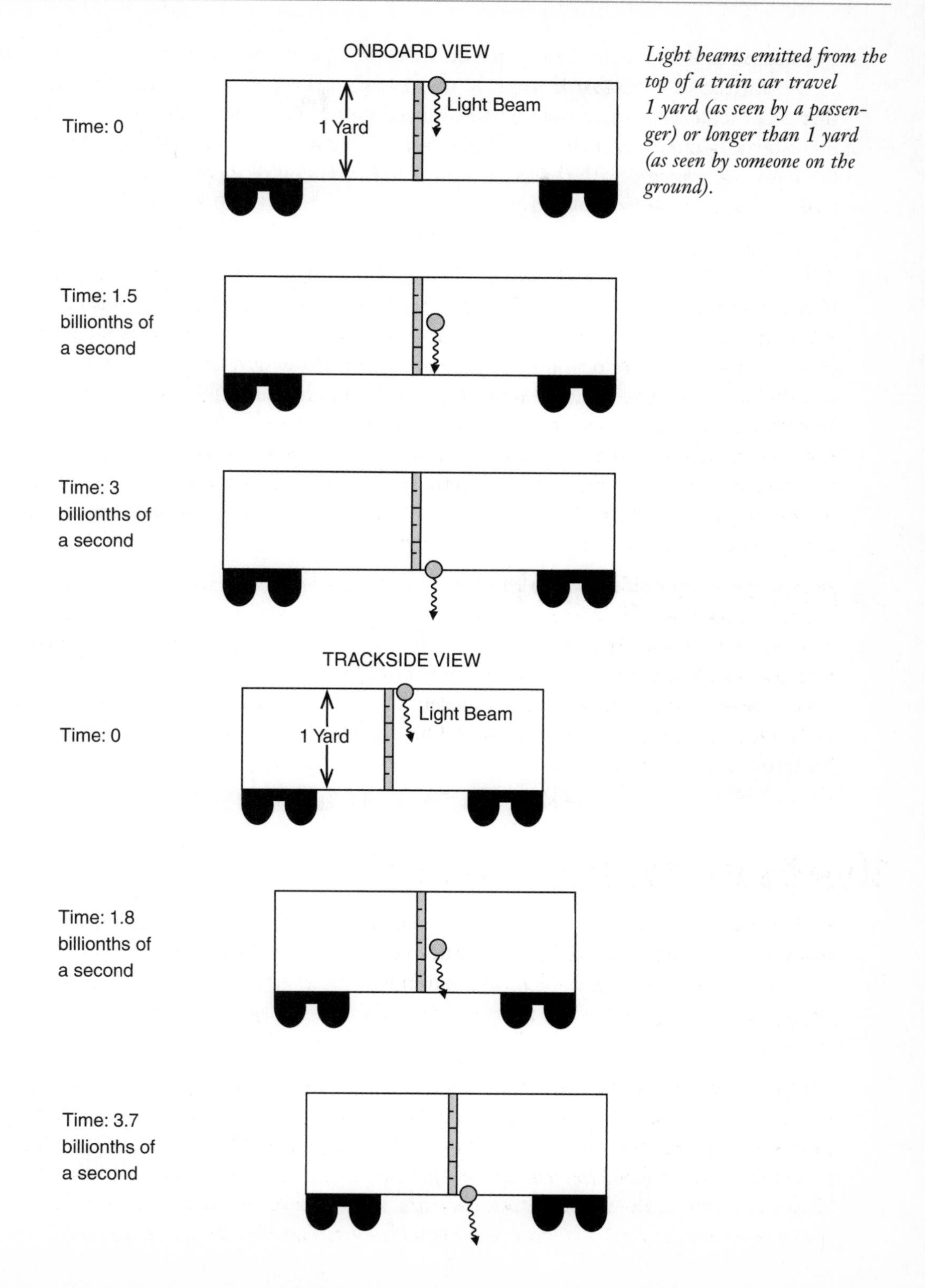

Light beams emitted from the top of a train car travel 1 yard (as seen by a passenger) or longer than 1 yard (as seen by someone on the ground).

Puzzled by this discrepancy, you pick up a pair of binoculars to look at my watch. You see it running slow compared to yours. But it's a Swiss watch, so the fault can't lie there. In fact, if you watch my body movements, they, too, seem a bit slow, as if I'd had a bit too much to drink. This situation is perfectly symmetrical. To me, you're the one moving, you're the one with the slow watch, and you're the one who needs to be scolded for drinking while doing experiments.

These conflicting impressions lead to no contradiction. We slice up the world into different moments, our clocks run at different rates, and, although I haven't gone into it, our distance measurements differ, too. But these effects, put together, ensure that we both see events occurring in the same cause-effect sequence, even if we disagree on the exact times and locations they take place. In the end, that's what really matters. Who really cares what the clock reads when the light hits the front of the train? Events still unfold in basically the same way for the two of us. Neither of us sees, for example, the front of the train emit a light beam that then travels backward and gets sucked up by the bulb. When the speed of light is the same for everyone, the order of events can never reverse. This fact will come up again and again throughout the book.

Space and Time, Unite!

No matter how compelling the logic, the idea of two simultaneous events becoming consecutive has a funny ring to it. It's funny in the same way your kitchen looks funny when you tilt your head. A flat countertop then looks sloped, and two jars of pickles at the same height no longer appear to be. The left-right direction gets jumbled with up-down.

Motion has an analogous effect. Two events at the same time no longer appear to be. The "direction" of time gets mixed up with a direction of space. Relativity theory freely trades time off against space. It treats them as two aspects of *spacetime*, just as the left-right and up-down directions are two aspects of space. Your sense of left-right differs from mine, depending on your viewing angle, and your sense of time differs from mine, depending on your velocity.

def•i•ni•tion

Spacetime is the union of space and time. Two people moving at different velocities see the same spacetime but divide it up differently into space and time.

Observable spacetime consists of four **dimensions**—four independent directions. Identifying the location of an event requires four numbers, known as **coordinates,** one for each of these dimensions.

When people say time is the fourth *dimension*, they don't just mean that it takes four numbers to pinpoint an event (three for space, such as latitude, longitude, and altitude, and one for what the clock says), but also that time can be traded off against the position within space. In fact, the *coordinates* you choose to identify locations are arbitrary: you need four of them, but as long as you're consistent about it, you can choose whichever four you want. Riding the train, I might choose the middle of the train car as my reference point, whereas you, standing in the station, might choose a signpost in the station. To relate these two sets of coordinates requires our taking time into account.

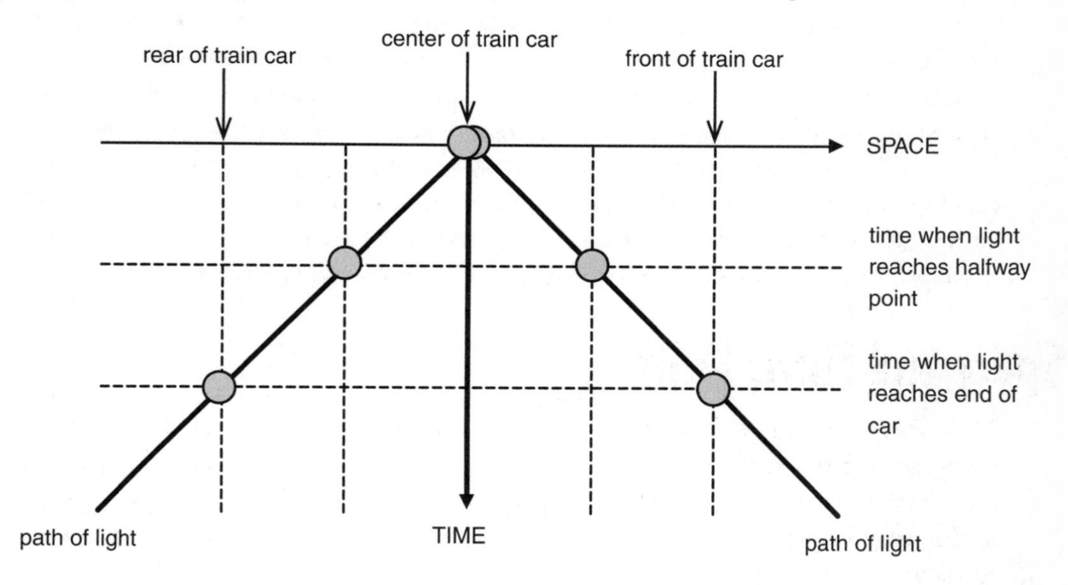

Space and time merge into spacetime, which we can depict using a map like this (as seen by the passenger on board the train).

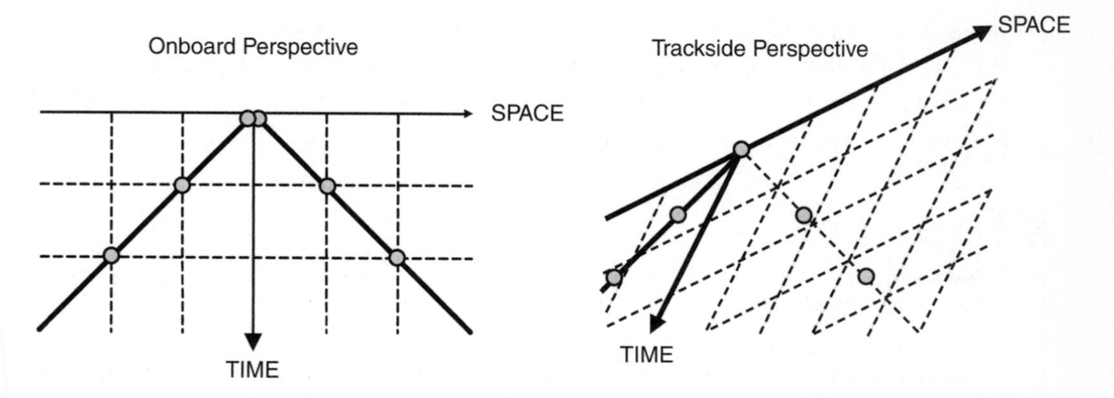

Two different observers slice up spacetime in two different ways depending on how fast they're moving.

Just because time is another dimension doesn't mean it's *just* another dimension. After all, you're not free to move in time as you are in space. Time plays the special role of distinguishing cause from effect. Seeing an effect precede its cause is analogous to tilting your head so much that the left-right and up-down directions don't just get jumbled, but swap places. The universality of the speed of light stops you from doing that.

Think Globally, Act Locally

If the speed of light is the same for all objects, then no object can ever catch up to light. If an object could, then relative to it, light would not be moving at all—contradicting the principle that the speed of light is the same no matter how we view it. A kind of governor starts to apply the brakes as we approach the speed of light. Each additional increment in speed requires a disproportionate increase in energy, and to reach the speed of light would take an infinite amount of energy. So although we call it the "speed of light" for historical reasons, it's really a universal speed limit. (To be precise, this speed limit is the speed of light in a vacuum. The speed of light in a material such as glass or water is somewhat slower.)

All Tangled Up

Will we ever be able to travel faster than light? Relativity theory says you can't outrun a light beam in a fair race when both run down the same track. But if you took a shortcut, you could beat light (see Chapter 10). I hasten to add that human technology isn't up to the task, and a quantum theory of gravity may close the loophole that allows this possibility.

This limit, though it has unexpected effects on time, cleans up some icky aspects of pre-Einsteinian physics. If nothing limited the speed of objects, an alien being could scoot across the universe, steal half our laundry, and scuttle back to its home world before we'd ever know. This would explain the eternal mystery of mismatched socks, but it would make a hash of scientific theories—events could happen seemingly without reason, and distance would become meaningless. Philosophers of physics have called these aliens "space invaders" in homage to the classic arcade game.

def•i•ni•tion

Locality is the concept that what happens in one place doesn't instantly affect another. Something moving at a finite speed must pass between the two places to carry the influence. The opposite is nonlocality.

The speed limit ensures that objects have to cross space and time to get from one place to another. They can't just jump. If the aliens want to steal our laundry, they'll have to work for it. This restriction is known as *locality*. In everyday language, locality means "a place," as in a site or neighborhood. Here it means something more abstract: the fact that an object can *have* a place, that it can plop down in a specific location distinct from all other locations. Because of locality, space acts as a sea separating objects from one another, and time is a sea separating events. The physicist John Archibald Wheeler, quoting bathroom graffiti from a restaurant in Austin, Texas, once wrote: "Time is nature's way to keep everything from happening all at once." Likewise, space is nature's way to keep everything from happening in the same place.

Mass and Energy

I've avoided equations throughout this book because I think I can explain the gist of physics without them, but I have to give you one equation or you'd ask for your money back. And, of course, I am talking about the famous $E = mc^2$. This little formula says that a certain amount of mass (denoted m) corresponds to a certain amount of energy (E), and the conversion factor is the speed of light (c) squared. Through this equation, relativity theory unites not only space and time but also the concepts of mass and energy.

In physics, energy is the property of an object responsible for bringing about change. It comes in various forms that can be interconverted, such as kinetic energy (associated with motion), thermal energy (associated with heat or, equivalently, the motion of molecules inside an object), and chemical energy (associated with the structure of molecules). An object can gain or lose energy only by transferring it to some other object like a hot potato; energy is never lost. For its part, mass is the property of an object that resists acceleration (or change in velocity) and responds to the force of gravity. Einstein's genius was to show that mass is, indeed, another form of energy.

As with much else in relativity theory, the formula arises from considering how different observers view a situation. Suppose you're back on your station platform, and I'm on my train, this time having some fun with Velcro-covered balls. I stand at the rear of the car, and my friend stands at the front; we throw the balls at each other, and they collide in the center of the car. Their speeds, being equal, cancel out. The balls stick together, fall straight to the floor, and lie there.

For sake of argument, suppose we throw them at nearly the speed of light. As long as we both throw them at the same speed, the end result is the same: the velocities cancel out and the stuck-together balls stop moving. Let's also suppose the train is moving at the same speed relative to the station at which we throw the balls.

What does it look like to you? My friend was throwing the ball against the motion of the train, so its net speed is zero. The ball seems to hover as the train rushes forward around it. My ball moves with the motion of the train, so it gets a boost. But here's the kicker: it's already going at nearly the speed of light. Because it can't exceed this limit, the boost has only a minimal effect. My ball flies forward, hits my friend's, and they stick together. The balls then continue moving forward with the train's speed.

The sticky balls stick together after they hit; to an observer on the ground, the stuck-together balls move almost as quickly as the single incoming ball does, suggesting the incoming ball is heavier than the one it hits.

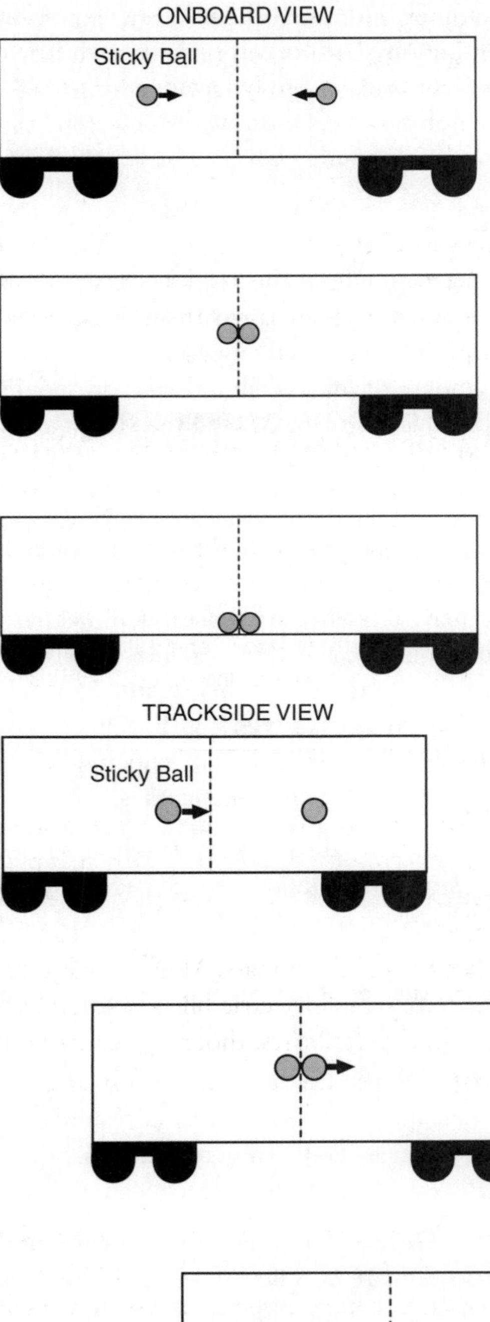

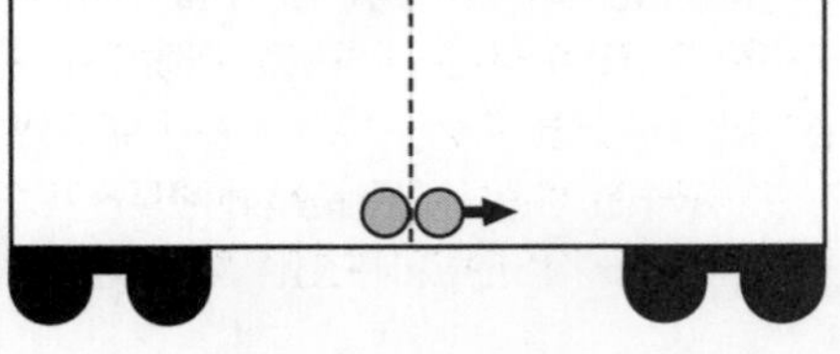

In short, from your viewpoint, my ball hardly seems to slow down at all. It's like a Hummer rear-ending a Mini Cooper. It picks up the little car on its front fender and barely notices. But what could possibly have turned my unassuming ball into such a monster? The only difference between my ball and my friend's is that it's moving. So the only explanation for its sudden heft can be that the energy of its motion is acting as mass.

Relativity allows mass and other forms of energy to be freely converted. The m in the $E = mc^2$ formula is the "rest" mass, the residual energy the particle has even when it doesn't appear to be moving. Rest mass is the energy intrinsic to the object—what's built into its structure.

Quantum Leap

Here's Einstein's original argument for $E = mc^2$. I'm back on my train. Just as I pass you in the station, I turn on a battery-operated lamp to illuminate both ends of the train car. To me, the situation is symmetrical. The light beams carry equal amounts of energy in both directions. To you, it's asymmetrical. The motion of the lamp stretches light moving in one direction and compresses it in the other, for a net increase in energy. To compensate, the lamp itself loses some kinetic energy. Because its velocity is fixed, its mass must decrease. In other words, some of the mass of the battery gets converted into light energy.

When an object changes its internal structure, it can change its rest mass, too, and the difference emerges in another form of energy. When you burn gasoline, the chemical waste products together weigh a little bit less than the gasoline originally did. This difference is converted to kinetic energy to run your car. Conversely, when you charge a battery, it gains ever so slightly in mass. Mass is such a concentrated form of energy that you'd rack up a considerable electric bill to make a noticeable difference. In particle accelerators and nuclear reactors, though, converting mass to energy and energy to mass is the whole point.

Atlas Shrugs

What if I don't have a nice smooth Swiss train to ride and instead have to make do with the herky-jerky New Jersey Transit? The train starts up so forcefully that I'm thrown back in my seat. When it screeches to a halt, I pitch forward. As it rounds a curve, I slam up against the window. For you, watching with pity from the station, there's no mystery: when the train starts, it takes force to get me to accelerate along with it, and when it brakes, my forward momentum carries me into the back of the next seat. But for me on board, it's as if an artificial gravity, acting in a horizontal direction, were being switched on and off. Maybe New Jersey Transit is more sophisticated than I give it credit for.

def•i•ni•tion

The **principle of equivalence** is the concept that acceleration produces an artificial gravity and, conversely, gravity is indistinguishable from acceleration. It's the basis of the general theory of relativity.

The principle of relativity still applies in these situations, but with a twist. The laws of physics are the same for everyone, as long as those laws include the effects of gravity. What you perceive as changes in motion, I perceive as a type of gravity. This idea, called the *principle of equivalence*, underpins the general theory of relativity. When Einstein formulated the theory, he started a trend of roping in new forces in order to ensure the laws of physics remain the same under ever-more-diverse circumstances (see Chapter 17).

The change in motion that gives the sensation of gravity is not the lurch of the train but the smack-up against the seat. As they say, it's not the fall that hurts; it's the sudden stop at the end. General relativity elevates this folk wisdom to a law of nature. For example, consider a skydiver. Normally we think of sitting still on the ground as the default mode and the skydiver who is deviating. The force of gravity reaches out from Earth to the skydiver and reels her in.

Relativity flips this reasoning on its head. Freefall is the default mode. Leaving aside air friction, the skydiver feels weightless—that is to say, she feels no gravity acting on her, artificial or otherwise. Only when the ground gets in the way does she feel any force, and even then, it's not gravity but the force of whatever stopped her. When she's sitting on the ground, the planet's stationary surface has to continue exerting a force on her to ensure she doesn't resume her freefall.

What determines the path the skydiver takes as she freefalls? Nothing is exerting any force on her, so it must be spacetime itself that guides her. Her velocity is continuously increasing, so it makes her perspective on spacetime a little more complicated than that of the train passenger moving at a steady speed. The skydiver's perspective changes from moment to moment during her plunge. If she draws gridlines indicating how she slices up space and time, those lines are not just tilted; they are curved.

Gravity, then, corresponds to bent spacetime. Bodies fall toward Earth's center because the mass of our planet distorts spacetime and makes the paths of freefalling bodies converge.

In short, spacetime comes alive. It curves, twists, grows, and shrinks. No longer content to carry the world on its back, it becomes an active participant in the drama of life. The shape of spacetime reacts to matter and energy. The density of material determines the curvature of spacetime, and the curvature, in turn, dictates what happens to the material. The effects are far richer than Newton's law of gravitation ever allowed for. For one thing, Einstein's theory predicts that all forms of energy produce gravity—including gravity itself. So gravity feeds on itself. If you pack enough energy into a small-enough space, gravity goes berserk and gives you a black hole (see Chapter 8).

By following the deceptively simple principle of relativity where it took him, Einstein created a theory that has lost none of its capacity to surprise physicists. His great ambition was to describe all of the physical world in purely geometrical terms. This dream is still unrealized. As I will discuss in the next several chapters, greater progress toward a comprehensive theory has come from a very different direction: quantum theory.

The Least You Need to Know

- The laws of physics are the same for everyone, and this equivalence changes our conception of space and time.
- Time is not fixed; the rate of its passage can change depending on speed.
- Space and time are two parts of the unified concept of spacetime.
- Forces take time to propagate from one place to another, a concept known as locality.
- Mass and energy can freely convert from one to the other, according to the equation $E = mc^2$.
- Spacetime is malleable, and variations in its shape produce what we perceive as gravity.

Chapter 4

The Quantum Revolution

In This Chapter

- Seven wonders of the quantum world
- Quantum theory in a shell game
- Wave action and types of particles
- How quantum effects get hidden

Quantum theory is such a luxuriant rainforest of a theory that it makes all the theories that came before it look like rock gardens. Prior to quantum theory, physicists really had no idea what matter is. They took the properties of matter as givens and ran into contradictions when they dug deeper. Quantum theory resolves those contradictions and greatly expands our view of nature. Although it takes a very different approach to describing nature than relativity does, the theories have an intriguing similarity: both acknowledge our view of the world is biased and try to take this bias into account.

Not as Weird as They Say

When scientists (and certainly nonscientists as well) talk about quantum theory, they generally start by describing it as irredeemably weird and difficult. Yet, although there's no doubt that quantum theory is a puzzle box, so were prequantum theories. To make things worse, those theories were puzzles also short of a few pieces.

They had serious trouble explaining atoms, for example. They predicted that each atom would be unique, creating an infinite number of chemical elements; that electrons would spiral in on atomic nuclei, with unhappy consequences; that the electrons themselves would explode under their own pent-up electrical forces; and that the light emitted by atoms would fry us with deadly radiation. Prequantum theories offered a grim view of a world bent on self-destruction.

Quantum theory explains why the world is at peace with itself. It says that quantities such as the energy of electrons within an atom can't take on just any old value; they come in discrete amounts, like the steps on a staircase. For instance, an electron might have 1, 2, or 3 units of energy only; it can't have intermediate values such as 0.707 or 1.414. Consequently, the electron can't spiral smoothly into the nucleus of an atom; instead it skips down the staircase steps from 3 to 2 to 1 and then stops or bounds back up. The lowest step corresponds to the nearest possible location to the nucleus. Only in the most extreme situations can an electron get shoved into the nucleus. Similar limitations apply to light and electrical forces, curbing their potential excesses.

In the Loop

Quantum mechanics is very much more than just a "theory"; it is a completely new way of looking at the world, involving a change in paradigm perhaps more radical than any other in the history of human thought.
—Anthony Leggett, University of Illinois at Urbana-Champaign

Quantum Leap

Isn't quantum theory just about teeny-tiny particles? Nope. Although its distinctive effects become easiest to see for subatomic particles, most physicists think it applies to everything, whatever its size. Whether you can see these effects or not depends less on a system's size than on its complexity. Experimenters have observed quantum behavior with the unaided eye.

Despite these insights and understandings, no one agrees on what the darned theory really means. Although it works mathematically and makes incredibly accurate predictions, it lacks the widely accepted conceptual footings that relativity theory has. Whereas Albert Einstein started with principles and then developed his equations, quantum physicists have had to work backward from the equations to the principles. For much of the twentieth century, they all but gave up.

In the past decade, though, a huge revival in interest in the meaning of quantum theory has occurred, driven partly by the needs of quantum gravity. I'll encapsulate some of that understanding here, but

I warn you upfront: this is the cutting edge of science, and there's hardly a thing I can say without guaranteeing my inbox will fill with irate messages from one researcher or another.

Seven Insights of Quantum Theory

Quantum theory introduces so many new ideas into human thought that even the experts are still processing them all. These are some of the most important insights, especially those that will prove important to understanding string theory, loop quantum gravity, and similar efforts to unify physics.

- **The world is digital.** Many essential properties of nature are discrete—that is, they come in distinct units, like bytes of computer memory or pixels on a screen. For instance, matter consists of particles. Particles' properties are often restricted to a limited number of allowed values. Even when properties can take on a continuous range of values, they are restricted in other ways.
- **To everything there is a reason.** Quantum theory is not magical or subjective. The state of a quantum object is completely well-defined and *deterministic.* The theory predicts how the state changes with time without any ambiguity. The notorious unpredictability of the theory arises only when we try to relate this state to our observations. A single quantum state can correspond to multiple positions, and when you go to look for the particle, you find it in one of these positions at random.

def•i•ni•tion

Determinism is the principle that the universe follows a specific course of behavior. Two identical situations lead to exactly the same outcome. The opposite is indeterminism, in which two identical situations lead to different outcomes at random.

- **Information is limited.** You can glean only so much information about a quantum particle. The particle is like a character in the computer game *The Sims.* When you set up the character, you have a certain number of personality points to allocate among various traits. So you have to make tradeoffs. A Sim can be nice but sloppy or neat but grouchy, but he can't be nice and neat at the same time. For a particle, you need to allocate your points among properties, such as position and velocity. If you know exactly where the particle is, you have no idea how fast it's going. If you then measure its velocity to multiple decimal places, you lose track of its position.

- **You can't paint a red spot on a particle.** Two particles with the same intrinsic properties are perfect clones. You can never be sure they haven't swapped places. You can't imagine putting a red spot on one of them to tell it apart, even for the sake of argument, because that would amount to extra information it has no place to store. This indistinguishableness governs the particles' behavior, although philosophers disagree on whether it's truly fundamental to their nature.

- **The whole is greater than the sum of the parts.** Many of the properties we ascribe to bodies are not actually inherent to them but describe their relationships with other bodies. If you look at a random-dot stereogram with one eye, it looks like meaningless static. If you use both eyes, though, you can see a pattern. Similarly, two *entangled* particles can have random properties when viewed separated, but when you look at them together, a pattern becomes evident. The pattern is more complex than you'd expect from the particles' own limited information-storage capacity. Either the particles are coordinating their appearance across the space that separates them (through a so-called nonlocal connection), or some other mysterious effect gives the illusion of such a link. The next section will discuss this further.

def•i•ni•tion

Entangled particles form a single joint system. The system has properties that we can't explain as properties of the individual particles in isolation.

- **Our view of the world is unavoidably biased.** One of the reasons we can't apprehend the quantum world fully is that we're part of that world. Each of us sees part of the world, part of the time. We can arrive at conclusions that are flatly contradictory. For me, a particle may not have a well-defined position; for you, it may not have a well-defined velocity. No amount of Kissingerlike diplomacy could reconcile us. The beauty of quantum theory, and the reason physicists think it captures something objective about the world, is that when we actually come together to compare notes (an interaction that is itself subject to quantum laws), we find they're entirely consistent.

- **Everything not forbidden is compulsory.** According to prequantum theories, a particle always has a definite location. Even if you don't know where it is, you can assume it's *somewhere*. If you prepare a number of particles in the same way and treat them the same, they'll all end up in the same place. According to quantum theory as most physicists understand it, that's not the case. When you can't tell where a particle is, you have to treat it as though it's everywhere it could possibly be—at once. If you prepare enough particles in the same way, they'll wind up in different locations at random. They'll go every place and do everything the theory does not strictly rule out.

In the Loop

I recall also being very struck that there were atoms in my body that were entangled inextricably with atoms in the bodies of every person I had ever touched.

—Lee Smolin, Perimeter Institute

The bottom line is that the world is made of some very interesting but inscrutable stuff. Quantum particles may be simple creatures, but they're squirmy and high-strung. You can't so much as look at them without sending them flouncing off. They have a much richer repertoire of behaviors than prequantum theories ever envisioned, although physicists disagree on what precisely makes them distinctive. Each of the seven aforementioned principles has been proposed at some point as the true essence of quantum theory.

In the Loop

The world is sensitive to our touch. It has a kind of "Zing!" that makes it fly off in ways that were not imaginable classically. The whole structure of quantum mechanics … may be nothing more than the optimal method of reasoning and processing information in the light of such a fundamental (wonderful) sensitivity.

—Christopher Fuchs, Perimeter Institute

The fact that the world resists complete understanding doesn't mean that knowledge is impossible or that anything goes—just that quantum theory is very specific about the kinds of information we can gather and what happens to the world in the process of gathering. Accepting these limits is the first step to expanding our knowledge.

The Games Quanta Play

Most physicists are pet-lovers, so it's a little awkward that the iconic example of quantum theory is the tale of Schrödinger's cat, a poor kitten that gets caught in a state of being both alive and dead per the "Everything not forbidden is compulsory" rule. But let's not go there. Instead, let's consider a couple of less morbid puzzles that reveal the novelties of the quantum world even better.

Bell's Shells

Imagine that you and your sister go to a carnival and come across an unemployed Ph.D. physicist running a shell game. You sit down to play, each of you with three shells. The gamester hides a ball underneath one shell in each of your groupings, then mixes the shells up, and you each try to choose the one shell from your three that has a ball hidden underneath it.

You and your sister play the game over and over, choosing shells at random. You quickly notice you have a 50–50 chance of finding a ball under any one shell, and your sister notices the same. Overall, the two of you have a 50–50 chance of seeing the same thing (either a ball or not) or different things (one sees a ball, the other doesn't). But something funny is going on. Coinciding with this randomness is a peculiar pattern. Whenever both of you pick up the same shell (left, right, or middle), you both see the same thing.

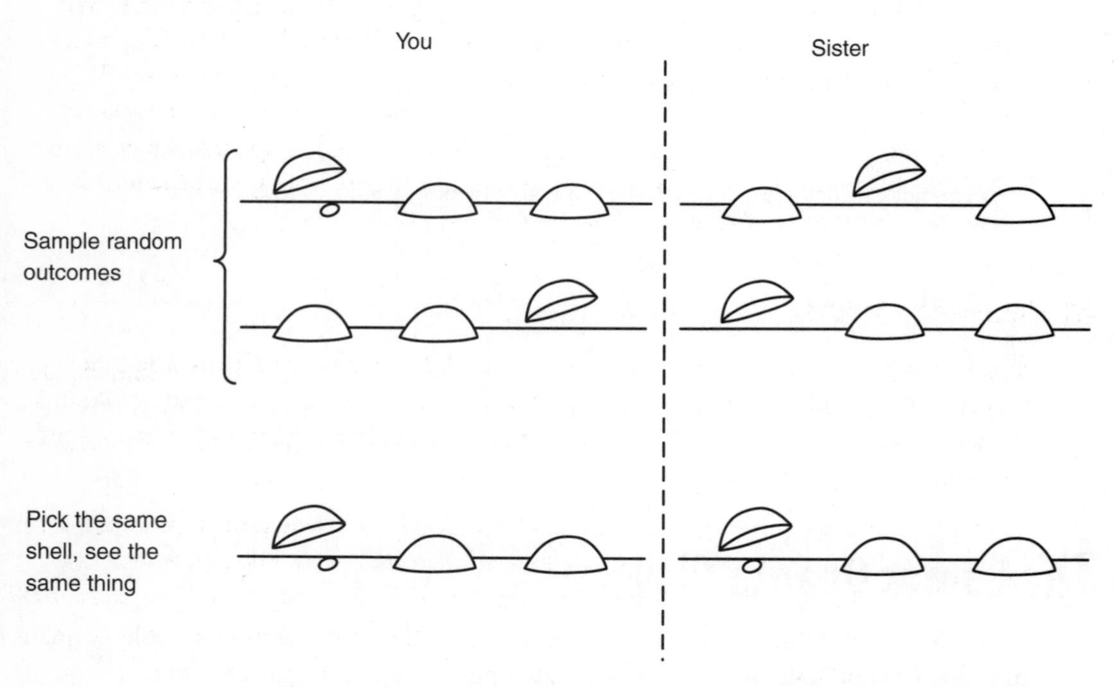

You and your sister pick up shells and see a ball, or not, at random—but if you both pick up the same shell, you see the same thing.

"Gotcha!" you say and announce you've figured out the trick. Because the gamester doesn't know in advance which shell you're going to choose, to ensure the above outcome, he must have set up your shells and your sister's with the same arrangement of balls each time. "Not so fast," your sister replies. Working through the probabilities in your explanation, there's a one-in-three chance you both choose the same shell, in which case you always see the same thing. The other two thirds of the time, you turn over different shells, and in half of these cases you see the same thing. All in all, there's a two-in-three chance that you both see either a ball or no ball. This probability contradicts the fact you have an even chance of seeing a ball.

So there's no way the gamester could have set up the balls in advance to produce the desired effect; he must be using some sleight of hand. Maybe the gamester watches the two of you choose shells and at that moment distracts your attention momentarily so he can fix the arrangement. To foil him, you and your sister make sure to pick up your shells at exactly the same time. And yet he still pulls off the trick. Are his hands moving faster than the speed of light?

It turns out the supposedly unemployed physicist works for Caltech and is just supplementing his academic salary. This cunning carnival game is equivalent to an actual experiment devised by the Irish physicist John Stewart Bell in the 1960s. The two sets of three shells correspond to two particles that we can measure in three ways. When we measure them in different ways, we get random answers, but when we measure them in the same way, we get the same answer. No out-and-out inconsistency arises, but when we ask what's going on behind the scenes, there's no way to explain it in purely prequantum terms.

An Even Odder Game

The following year, you and your sister return to the carnival and track down the gamester. He greets you with a devious smile and presents you with a new game consisting of only four shells. Each person must pick up either the left or the right shell, which could have a ball under it or not.

After you play a few times, you notice a pattern. Whenever one person picks up the left shell and sees a ball, and the other picks up the right, the second person sees a ball, too. Whenever both of you pick up the right shell, at most one of you sees a ball. To ensure these outcomes, the gamester would have had to arrange the shells in one of the following five ways:

- No balls under any shell.
- A ball under your right shell and none under the other three.
- A ball under your sister's right shell and none under the other three.

- A ball under your right shell and your sister's left shell, and no balls under the other two.
- A ball under your left shell and your sister's right shell, and no balls under the other two.

There are 11 other arrangements for the balls that break the pattern. I admit this is hard to juggle in your mind, so I've drawn out these 16 possible arrangements. Only #1, #2, #5, #7, and #10 satisfy the above rules. Try different combinations of shells, and you'll see that the other options break the pattern.

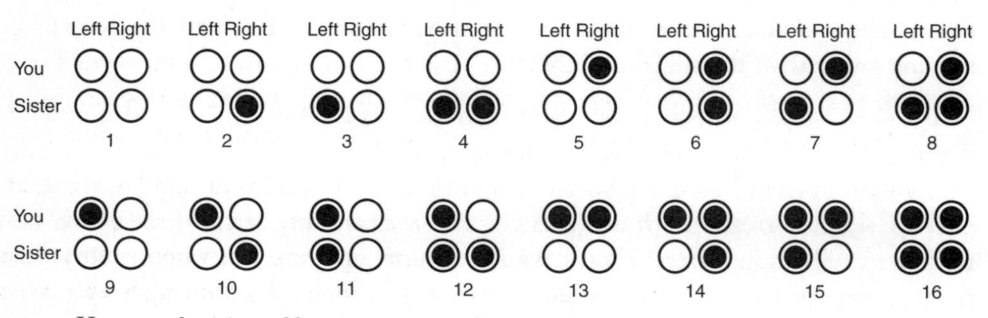

Here are the 16 possible arrangements of two balls under four shells.

Just when you thought you had it figured out, something comes out of left field. Both of you pick up the left shell and see a ball. That's not possible in any of the five allowed configurations. There's no way the gamester could have prearranged the balls to ensure it. And yet quantum theory allows it.

This is just a variant of Bell's experiment in which the two sets of two shells correspond to two particles that we can measure in two ways. The beauty of this variant is that you don't need to keep track of probabilities. Once you figure out the rules, a single counterexample breaks them, and the only way to explain what happens is to use quantum theory.

Figuring Out the Trick

According to prequantum theories, objects first acquire specific properties and then, optionally, coordinate those properties. But quantum theory allows for the reversed sequence. Objects can become coordinated and only then acquire specific properties. This is what I meant earlier when I said that relationships between objects can embody information that the individual objects do not.

That is to say, you'd ordinarily assume that, when you first look at the gamester's shells, they either have a ball under them or not. When you choose, the gamester must use some sleight of hand to ensure the shells match in the way you observe. But by using quantum particles, he can reverse the sequence. He first coordinates the shells—by creating a matched pair of particles—and then lets the particles decide whether a ball appears. The gamester is as clueless as you are about whether you'll see a ball under a shell. All he can be sure of is the overall pattern you and your sister will see.

Most physicists conclude that a nonlocal link connects the particles—that is, some means for them to communicate directly without having to send a radio signal, laser beam, smoke signal, carrier pigeon, or other middleman. The link allows the particles to coordinate their behavior on the fly. Yet the link can't transmit a message in the conventional sense. Although it lets particles coordinate their behavior, they, not you, decide what this behavior will be, so there's no way to encode data on them.

Hey wait, wasn't the special theory of relativity supposed to have ruled out instantaneous, nonlocal links? Yes, it was. Physicists and philosophers have yet to fully reconcile the locality of special relativity with the nonlocality of quantum theory. The saving grace of these links is that because we can't encode data on them, we can't use them to do paradoxical things such as invert cause and effect.

Still, not all physicists and philosophers accept nonlocal links. Some think the argument for these links has a flaw—namely, it ignores how our views of the world are unavoidably biased. According to the standard argument, before you pick up a shell in the carnival game, the shells are in the purgatory of both having and not having a ball. When you pick one up, it chooses. You assume that your sister has seen a definite result, too, even if you don't know what it is yet. From this you conclude that something (a nonlocal link) must have transmitted the outcome from your side to hers in order to make sure her shells will follow the right pattern.

What might be more proper to assume is that your sister's shells, and indeed your sister herself, remain in purgatory until you actually ask her what happened. Until that moment, you think you have a definite result and your sister doesn't, and she thinks she has a definite result and you don't. You have two incompatible views of the same situation—and that's fine. All that matters is that when you come together to ask each other what you've seen, you give the same answer—and you know you will, because the gamester prepared particles that would yield the same results, even if he didn't know what those results would be. We don't need any nonlocal links.

> **In the Loop**
>
> For those interested in the fundamental structure of the physical world, the experimental verification of Bell's inequality constitutes the most significant event of the last half-century.
>
> —Tim Maudlin, Rutgers University

Are you thoroughly confused yet? If it makes you feel any better, these ideas confuse physicists, too. They are among the deepest concepts in all of physical science. The concept of a nonlocal link will come up again in later chapters, so we'll have a chance to explore it from different angles. For now, the most important thing to take away is the appreciation that reality is much more interesting than prequantum theory made it out to be.

Wave of Chance

In Chapter 3, I described the ordeal of using New Jersey Transit trains, but the roads in the Garden State are no better. I can come around a bend or out of a tunnel and have to make a split-second decision—go left or right! One lane will take me home, the other onto a labyrinth of potholes taking me miles out of my way. I can't look to the signposts for help, as they are designed to send me in circles. So I must quickly decide, but wouldn't it be great to be able to check out both routes before committing to one? A quantum particle has that power.

If a quantum particle comes to a fork in the road, the particle takes it.

(Courtesy of Dan Moloney, Threepwood@b3ta)

A particle takes all the paths available to it, and those multiple trajectories determine the range of locations that the particle can get to. This is what I meant earlier when I said that everything not forbidden is compulsory. By allowing particles to take two paths at once, the theory creates a continuous range of possibilities. The particle can stay mostly on the left road, somewhat on the left, somewhat on the right, or mostly on the right. Up to the point of our measurement, the particle avoids sudden, irrevocable choices. This is essential to the determinism of the theory. In the quantum world, "maybe" means "yes."

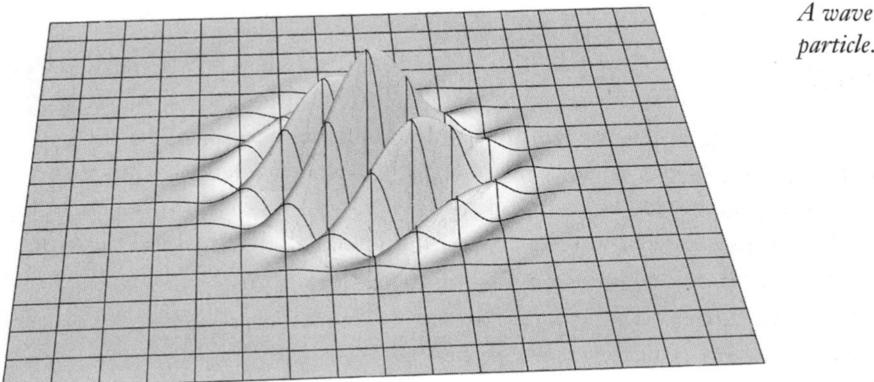

A wave is describing a particle.

A very familiar process has the same ability—namely, wave motion. When a wave reaches a fork in the road, the wave takes it; the wave splits into two. When the road merges back together, so do the waves, though they're never quite the same again. The fact it split lets the wave reach regions it otherwise couldn't.

So waves are a natural way to describe quantum particles. The wave associated with a particle ripples around in a completely predictable way. When you actually go to look for the particle, you find it at one and only one location, chosen at random from the locations spanned by the wave. The height or depth of the wave at a location tells you the probability that you'll find the particle there. A tall, skinny wave means you'll almost certainly find the particle in that place. A low, broad wave means it could be almost anywhere.

The probability depends on the distance of the wave from the horizontal centerline. A wave peak with a height of one unit looks the same as a dip with a depth of one unit. Whether the wave sticks up or drops down becomes significant only when waves overlap. A negative height can offset a positive one, reducing or even eliminating the probability of finding a particle in a given location.

All Tangled Up

What exactly is this wave whereof I speak? That's a good question. Some physicists and philosophers think the wave in the theory corresponds to a real wave out there in the world, perhaps a force field that guides the particle. Other researchers consider the wave merely a theoretical bookkeeping device, which distills the patterns in past measurements to make guesses about future ones. For them, whatever is out there in the world is not really a wave but some other kind of thing yet to be figured out.

One of the nice things about the wave description is that it explains how a particle can have both continuous and discrete properties. If you pluck a guitar string, it doesn't play any old note; it always plays a certain note and its harmonics. This is because the string is pinned down at both ends, which determines the wavelength (hence the pitch) of the sound. A trapped particle works the same way. In its case, the wavelength corresponds to the momentum of the particle. This and the other wavy qualities will become crucial in the coming chapters as I delve into the behavior of particles.

The Clone Armies

As anyone who's ever fallen in love will attest, having similar personalities can either be good or bad for a relationship. You might bond easily, complete each other's sentences, and win *The Newlywed Game*, or you might just drive each other nuts. So it is with quantum particles. Particles of a given type are all perfectly indistinguishable, but they have two ways to be indistinguishable: either they fall into an embrace or they hold one another at arm's length. We call the embracing kind *bosons* and the standoffish kind *fermions*, the terms deriving from the names of the physicists who explained them.

def•i•ni•tion

Bosons (slant-rhymes with "morons") are particles that don't object to clumping with identical copies of themselves. Their role in the world is to transmit forces.

Fermions are particles that tend to stay clear of identical copies of themselves. They are the building blocks of matter.

Quantum Leap

The boson's huggableness lets you put a huge number of identical bosons all together, of which the pure light of a laser beam is an example. When you try to bring fermions together, on the other hand, they resist. Their mutual aversion is a force to be reckoned with. The core of the planet Jupiter, for example, is filled with fermions packed together like sardines. Their resistance to tighter packing holds up the entire weight of the giant planet.

The reason there are two ways to be indistinguishable has to do with how quantum waves work. Since a peak of a quantum wave looks the same as a dip of the same size, flipping a wave upside-down has no outwardly visible effect. So if you swap two particles, the wave representing them could either flip or not, and no one will be able to tell the difference. The pair will still look the same before and after. Particles that don't flip when interchanged are bosons, and ones that do are fermions. Although fermions' flip-flopping doesn't affect their outward appearance, it does have an indirect effect—namely, it is what makes them so antisocial. If a peak marks the position of one fermion in a pair, then a dip marks the position of the other, and bringing them together cancels them out.

The geometry of particle-swapping bears an uncanny resemblance to salsa dancing. Switching places with your partner is a common move in salsa, and one thing you quickly learn is that both you and your partner need to do a half-turn to remain facing each other. If one person turns less, the other has to turn more, so that you do a full turn between the two of you. The same goes for particles. From their perspective, an interchange is really just a 360-degree turn. We normally think of a 360-degree turn as equal to no turn at all—it leaves dancers looking just like they did before (though maybe a bit dizzier). This intuition applies to bosons: after a full turn, they're back where they started.

Fermions, however, flip over and require a second full turn to bring them back where they started. The geometric reasons are actually very deep and have to do with how particles relate to their surroundings. Salsa dancers experience this when they turn while holding both of their partner's hands, in which case a full turn does *not* bring them back where they started, but instead tangles up their arms. A second full turn, combined with some clever arm movement, releases them. (For a movie that demonstrates this idea, see Appendix B.)

def•i•ni•tion

Spin is a quantum property related to rotational motion. It determines whether a particle is a boson or a fermion.

Accordingly, physicists distinguish bosons and fermions by a property called quantum *spin*. Each particle has a fixed value of spin depending on how many times you need to turn it to restore it to its original condition. Spin plays a crucial role in the development of string theory.

The Undocumented Feature

Plenty of features of quantum theory take some getting used to but aren't really mysteries; they're merely unfamiliar. If it's mystery you want, it lies at the boundary between the predictable theory and the random output. Quantum theory says a particle can be found in any number of locations, but we find it in one. What singles out that possibility among all the possibilities? Not only does the theory offer no explanation, but it also says that an explanation is impossible. The particle just does what it does, and that's that.

Will a unified theory of physics fill in the blanks, or will the selection process remain beyond rational understanding? No one yet knows. However, physicists have made progress on part of the puzzle. They may not know why we find particles in a given place, but at least they know why we find them in *one* place rather than many places at once. To be more specific, if the particle takes a certain path, the fact it might have taken another path shouldn't affect it. The paths need to break off their relationship and act independently.

They do so using a technique familiar to generations of disillusioned lovers: they start a new relationship. Suppose you have a particle caught between two paths and you bring it into contact with another particle. The particles establish a relationship, which dilutes the original one between the paths. As more particles join in, the original relationship gets so diffused that it effectively ends. The paths become independent alternatives.

def•i•ni•tion

Decoherence is the process that causes quantum particles to lose their distinctively quantum behavior—for example, in the process of making a measurement.

In terms of waves, what happens is that the original particle's waves split and take two paths, but those new particles butt in, throw them out of alignment, and prevent them from ever recombining, leaving them to go their separate ways.

An important aspect of this process, known as *decoherence*, is that the relationship is never truly lost but just spread out among multiple particles. It's similar to what happens with energy. If you throw a Superball off a three-story building, it will bounce up and down and eventually come to rest. But the energy of its motion hasn't disappeared. It's gone into the vibrations of the atoms in the ground and the ball—in a word, heat.

All Tangled Up

Is reality subjective? Many older books on quantum theory imply as much; some suggest that our observations create reality. In this view, quantum systems are caught in a purgatory, not knowing which of their multiple possibilities to choose until, by measuring them, we force them to decide. But in the modern view of decoherence, nothing is special about our measurements. Natural interactions have the same effect.

One of the lessons of decoherence is that although we think of the quantum world as strange and unfamiliar, it is *we* who are strange. Our everyday world is just the corner of quantum-land where decoherence has run to completion. We are only children playing on the beach, while the vast ocean of quantum reality lies undiscovered before us.

The Least You Need to Know

- Quantum theory is the physicist's theoretical framework for understanding matter.
- To a large extent, the quantum world is pixilated, divided into units (particles) with discrete properties.
- Objects limit what we can learn about them, leaving an element of randomness in our observations and forcing tradeoffs in measuring quantities such as position and velocity.
- Quantum theory blurs the distinction between possible and real—events that might have happened have observable effects even if they didn't come to pass.
- Far-flung objects appear to be linked together in a way that still perplexes physicists.
- Particles come in two basic types, fermions or bosons, depending on how they treat particles identical to themselves.

Chapter 5

The Standard Model of Particles

In This Chapter

- Types of particles
- Types of forces
- Quantum fields of dreams
- The reality of virtual particles
- Why is the weak nuclear force weak?

The generation of physicists who introduced such evocative terms as "quark" and "black hole" must have run out of name ideas when it came to the *Standard Model.* The name makes it sound like a yellowish Chevette parked in a long row of identical yellowish Chevettes. In fact, the model is more like James Bond's Aston Martin DB5—a classy, one-of-a-kind creation with a button for almost every contingency. The Standard Model explains more aspects of the physical world than any other theoretical framework in the history of science, but it raises almost as many questions as it answers.

Zen and the Art of Particle Physics

When I was a kid, I loved taking my family's cassette tape players apart, although my family felt somewhat differently about it. A simple exterior—just a few buttons and dials—hid a complex interior, full of spinning wheels and wires. Each of those parts was simple, even boring, but how they were put together made them interesting. A 10-year-old had a reasonable shot at fixing it if it broke. In contrast, taking apart an iPod isn't nearly as satisfying. I don't learn a thing because the simple exterior hides a simple interior. All the complexity is packed into a couple of black chips, which I can't take apart without specialized equipment, let alone fix.

Fortunately for particle physicists, nature seems to work more like the cassette player than the iPod. We can take it apart; then take apart the pieces; and then take apart the pieces inside the pieces all the way down, until we disassemble atoms and atomic nuclei into *elementary particles*—at which point we have reached something too simple to require disassembly.

def•i•ni•tion

An **elementary particle** is the most basic building block of nature.

The **Standard Model** of particle physics is the current explanation for the composition of matter and the workings of electromagnetism, the strong nuclear force, and the weak nuclear force.

Physicists and philosophers of physics have long debated how to define an elementary particle, but it boils down to this: it's an object that's as boring as it can possibly be. It has no inner workings; it looks exactly like all other particles of its type; and it possesses only a few intrinsic properties such as mass and electric charge. String theory says an elementary particle is a string at heart, but even if so, it's just a single wriggling strand rather than an elaborate system of moving parts. For most purposes, you can think of an elementary particle as a midget billiard ball: uniform, featureless, and predictable.

Yet the utter monotony of particles means that the workings of nature are laid out before you, rather than stuck inside some black chip you can't open. The endless diversity of the world we see is built up out of just 12 types of matter particles and their variants. They interact with one another by firing off force-transmitting particles drawn from a second set of 12.

There may be additional particles not yet discovered, but these two sets suffice for nearly all purposes. The only known scientific phenomenon they don't explain is gravity. In fact, just three of the matter particles and one of the force-transmitting particles are enough to reproduce the vast bulk of nature. The rest rear their heads only in exotic settings, although their existence seems to be necessary to make the whole apparatus click together.

The Standard Model is the descriptive framework for these particles and interactions. Theorists use the term "model" the way hobbyists use "model" train set—a way to understand how the real thing is put together by reproducing it in a more manageable form. The Standard Model unifies quantum principles and Einstein's special theory of relativity, and new phenomena, not present in either of those two theories on its own, arise.

A Tale of Two Particles

Particle physicists might not admit to playing favorites among the particles, but two of them really are special. The electron is the official particle of the Information Age. Its motion in wires gives us electricity; its motion in radio antennas gives us loud cell phone conservations. In atoms, the nucleus sits still like a parent chaperoning a birthday party while the surrounding electrons jump up and down (absorbing or emitting light in the process) and hopscotch among atoms (binding them together into molecules).

The other special particle is the photon, the agent of all electrical and magnetic phenomena, most notably that of light. The photon is the indivisible unit of energy within a light wave. It's a tiny wave in its own right, and by overlapping enough of them, we get a light wave we can see with the naked eye. Whenever an electron or any other electrically charged particle changes speed or direction, it fires off a photon to communicate its new status to all concerned.

The electron and photon are the archetypes of the two classes of particles in the Standard Model. The electron is a particle of matter, the photon a particle of force. In the jargon, the electron is a fermion and the photon a boson (see Chapter 4). You can think of fermions and bosons as the nouns and verbs of the Standard Model. Just as grammar demands two parts of speech, so does the natural world. To build up intricate objects, you need particles that resist clumping rather than merging into formless puddles of jelly. Fermions fit the bill. But the very fact that fermions resist clumping means they can't work together to transmit forces. That job falls to bosons.

Particles of Matter

The fermions of the Standard Model come in various types:

- **Quarks versus leptons.** Quarks are nuclear particles; they bind together to make the protons and neutrons inside atomic nuclei. Leptons, which include the electron, don't hook up in this way.
- **Red, green, blue.** Each quark comes in three versions named after colors although they have nothing to do with ordinary color. Quark color is a property analogous to electric charge. Leptons have no color.
- **Left-handed versus right-handed.** All fermions have two distinct identities that are mirror images of each other.
- **Doublets.** This is a fancy word for "buddy system" in which the left-handed fermions pair up with each other. For example, the electron pairs with a particle called the electron neutrino.
- **Generations.** The doublets come in three progressively heavier groups, called generations. The particles that constitute atoms, the up quark, down quark, and electron, are all drawn from the lightest generation.
- **Matter versus antimatter.** Each type of fermion has an evil twin, an antiparticle. It has the same mass as the standard particle but the opposite electric charge and color. The electron's antiparticle is known as the positron. If the two meet, they annihilate in a burst of energy. The known universe is made predominately of matter.

The question of how all these types are related has driven much of the progress of physics in the past few decades.

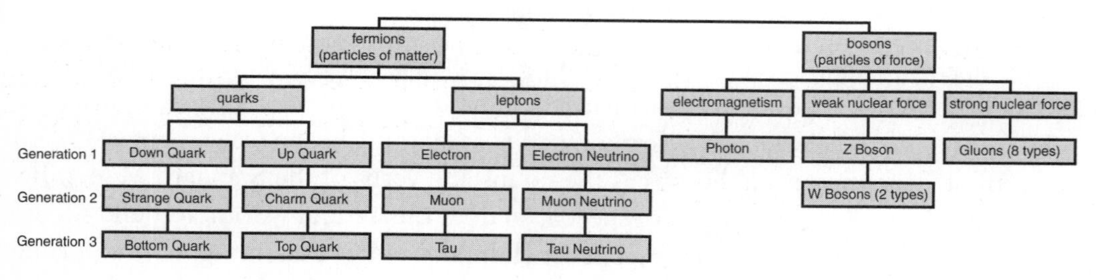

Categories of left-handed elementary particles (omitting antimatter and colored particles for simplicity).

Quantum Leap

Almost all particles have a limited lifespan; they are born, pay taxes, and die—recycling their energy back into the universe for new particles to form and continue the cycle. When a particle decays, it splits into one or more lighter particles and divvies up its properties among them. A few types of particles have no possible heirs, so they can't die. For instance, the electron is immortal because there is no lighter particle that could inherit its electric charge. The only known way to get rid of an electron is to slam it into an antielectron, or positron. The same goes for the up quark.

Particles of Force

Physicists know of four ways that particles cling to, push away, or flirt with one another. The Standard Model explains three of them:

- **Electromagnetism.** This is the most prominent force in everyday life. It drives electricity and magnetism, produces chemical reactions, generates light, and keeps buildings and mountains from collapsing under their own weight. To generate or feel the force, a particle must be electrically charged. Oppositely charged particles attract; identically charged ones repel.
- **Weak nuclear force.** This force is more obscure. It has a short range, a scant 10^{-17} meter, well inside the nucleus of an atom, so we don't normally feel it. Its main effect is to convert particles from one member of a doublet to the other—for instance, an electron to an electron neutrino or vice versa. In so doing, it can cause atoms to decay as surely as substituting a plank of balsa wood for a steel beam could bring down a building. To generate or feel the weak force, a particle must have "weak charge," a property that is similar to but distinct from electric charge. The two members of a doublet correspond to opposite values of this charge.
- **Strong nuclear force.** This is the muscleman of atomic nuclei. The protons inside nuclei are all positively charged, so they'd repel one another if something didn't hold them together. That something is the strong nuclear force. To generate or feel the force, a particle must have quark color, either red, green, or blue. Leptons are colorless, so the strong force doesn't affect them. Like the weak force, it has a very limited range.

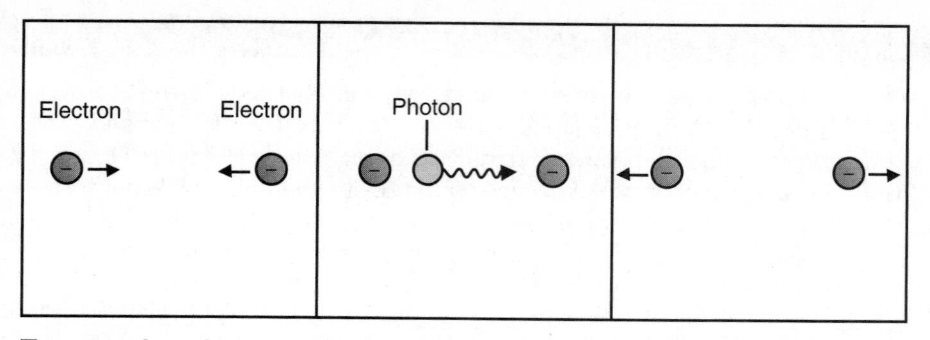

Two approaching electrons repulse each other by firing off a photon (it doesn't matter which electron fires first).

When particles exert forces on one another, the force must have some way of bridging the gap between them. That's where the bosons come in. They act as middlemen. When two electrons repel each other electrically, a photon shuttles between them. The nuclear forces involve larger contingents of bosons because these forces not only attract or repel particles but can also convert them to a different type. The weak nuclear force involves the so-called W boson and Z boson, and the strong nuclear force involves one of eight gluons, named because they glue the quarks in atomic nuclei together.

The Four Forces of Nature

Force	Intrinsic Strength	Typical Range	Carrier	Relevant Particle Property
Strong nuclear force	0.1	10^{-15} meter	Gluons	Quark color
Weak nuclear force	0.03	10^{-17} meter	W and Z bosons	Weak charge
Electro-magnetism	0.007	Infinite	Photon	Electric charge
Gravitation (not included in Standard Model)	10^{-39}	Infinite	Graviton(?)	Mass and energy

Particles and Fields

One quality that makes particles boring yet interesting is that every particle of a given type behaves in the same way. They may have different amounts of energy, but their intrinsic properties are identical. It's as though they were all cut from the same cloth. In the Standard Model, they are.

The cloth is a *field*, a substance that fills space like a fog. Like force fields in science-fiction—the gently shimmering walls that zap objects you throw into them—the fields of the Standard Model stretch across space and exert forces. The magnetic field is the best known example. We don't see it, but it's all around us, and if we throw down some iron filings, we can see its shape. Another common example is the electric field, which causes electric sparks. In fact, the electric and magnetic fields are two aspects of the same field, the electromagnetic field.

def•i•ni•tion

A **field** in fundamental physics is a substance that fills space like a fog. Describing it requires one or more numbers at each point in space. In quantum theory, a ripple in a field equals a particle.

Often we think of fields as coming from some object, like a magnet, but the Standard Model reverses the thinking; fields are fundamental, and particles come from them. Whenever a field gains some energy, it starts quivering like the surface of a pond, and these undulations correspond to particles. The amount of energy comes in discrete units, like M&Ms. You can't eat half an M&M (at least not if you are a true connoisseur), and you can't have half a unit of energy; it's either one or zero. One unit equals one particle.

Each type of particle has its own field, and the photon is the energy unit of the electromagnetic field. Photons are created or destroyed whenever energy is poured into or drained from this field. The electron is the unit of the electron field. The reason that every electron looks the same is that they're all generated by the same field, which stretches throughout the entire universe. The field is permanent, even as individual electrons come and go.

Antiparticles make sense in the field picture, too. If the particle corresponds to a traveling bump in the field, an antiparticle corresponds to a depression. When a bump and a pit meet, they cancel each other out, accounting for why particles and antiparticles annihilate.

Virtual Reality

Not only do fields explain the indistinguishability of particles, but they can also do things you might not expect. For instance, suppose you suck all the energy out of a field that you can. Empty is empty, you might think. But since when is modern physics so straightforward?

Even in a vacuum, the field is still there; you can't get rid of it completely. And it's always doing *something*. Like a pond surface, it's never perfectly still because stillness would imply an exactness of behavior that is alien to the quantum world. The field's irrepressible undulations generate particles. They die back down too fast for you to spot them individually, but their collective effects linger. Physicists call them "virtual" to signal that you can't observe them directly. They make the vacuum act a bit like a material in its own right, chock-full of particles even in the emptiest region of deep space.

All Tangled Up

Could the irrepressible energy of quantum fields solve the nation's energy problems? As with antimatter, science-fiction writers have speculated about tapping the energy that fields have even when they're supposed to be empty. But the observed density of such energy is very small. Moreover, drawing from it would leave a region of space with less than zero energy, a debt that would have to be repaid quickly—with interest (see Chapter 10). In the end, we'd be worse off.

Virtual particles also play a central role in particle interactions. Suppose two electrons come straight toward each other, repulse each other, and move apart again. The repulsive force is conveyed by a photon passing between them. Each electron changes direction but not speed. Its momentum reverses, but its energy (related to speed) doesn't change. The photon supplies the difference in momentum, but it contains no energy because the electrons don't need any.

From a prequantum point of view, that's impossible. The photon's momentum implies some energy, yet the photon isn't supposed to be carrying any energy. Quantum theory, though, allows this behavior as long as the photon lasts too fleetingly to be detected directly—that is, as long as it's virtual.

All Tangled Up

Do virtual particles violate the law of energy conservation? Books on physics commonly state virtual particles conjure up energy out of nothing for a fleeting moment. This description is misleading. Energy can't be created, period. That's a bedrock principle of all modern theories of physics. What makes virtual particles special is that they violate the conventional formula relating momentum to energy, which was never sacrosanct anyway.

All fermions and bosons can appear in both a real or virtual form. The adjective "virtual" is a little unfair because virtual particles are no less authentic than real particles are. Physicists may not be able to see them directly but can detect them en masse. Virtual particles also play a starring role in the unification of physics. You'll see them throughout the book.

The Weird Nuclear Force

If they were giving out awards for the weirdest force, the weak nuclear force would surely win. Electromagnetism is fairly simple, and gravity and the strong nuclear force, once you get past their apparent complexity, are elegant. The weak force, on the other hand, is a problem child.

- **It's lazy.** The "weak" force isn't weak so much as slothful. It's actually about as strong as electromagnetism, but it needs a swift kick to get it to do anything. Creating W and Z bosons, unlike photons and gluons, requires a certain minimum threshold of energy, so they spring into action only under sufficiently energetic circumstances that arise only over short distances.
- **It's klutzy.** Once the bosons swing into action, they act like bulls in a china shop. If untamed, they become intensely interacting at energies not far above the threshold for creating them, with all sorts of awkward effects.
- **It's lopsided.** The weak force isn't mirror-symmetric. It acts only on left-handed particles; the righties go scot-free. For antiparticles, it's the other way around: only the righties engage with the weak force. In other words, when it comes to the weak force, particles and antiparticles are like vampires: they don't cast an image in a mirror.

- **It's biased.** Particles from different generations differ only by mass—they have the same electric charges, weak charges, and quark colors. Accordingly, electromagnetism and the strong force treat them all the same but not so the weak force. It drives seemingly equivalent reactions at different rates. It even treats antiparticles slightly differently from particles.

It turns out these issues are related, the common thread being the concept of mass. The energy threshold required to create W and Z bosons is their mass; left- and right-handed particles are united by a common mass; and mass is what distinguishes the particle generations. So the idiosyncrasies of the weak force are bound up with the broader question of why particles have mass at all.

def•i•ni•tion

The **Higgs field** is a type of field thought to be responsible for giving elementary particles their mass. The Standard Model has several of them.

Before the Standard Model, physicists had no good answer to that question. Mass was just one of those things that particles had. The Standard Model demanded a deeper answer, largely because of the peculiarities of the weak force. The answer was the *Higgs field*, named after one of the scientists who proposed it. It's an additional type of quantum field on top of all the others.

To create a W or Z boson, you can't just pump energy into the boson's own field; you need to add some to the Higgs field as well. This sets a threshold for boson creation. So that's the W boson's excuse for being lazy. If you were surrounded by the Higgs field, you'd be lazy, too. Er, actually you are. The elementary particles in your body have to wade through the Higgs field whenever their velocity changes. The Higgs makes it harder to accelerate particles, an effect we perceive as mass.

All Tangled Up

Is the Higgs particle the godlike source of all mass in the universe, as some books have said? Leaving aside the questionable theology, that's an exaggeration. The Higgs can account for the mass of elementary particles but not of composite particles, such as protons and neutrons. The energy that binds these particles together also acts as mass, so they are much heavier than the sum of their elementary particles.

If you pump enough energy into the Higgs field, you create independent Higgs particles and liberate ordinary particles to act as though they had no mass at all. At that point, the weak force sheds its lazy habits and learns to treat particles evenhandedly. The distinction between electrons and electron neutrinos goes away.

In short, the weak force begins to behave much like electromagnetism. These two forces bear more than a casual resemblance. The Standard Model holds that they merge into a single set of forces, known as the *electroweak forces*. Photons mingle with the W and Z bosons, and electric charge with weak charge.

def•i•ni•tion

The **electroweak forces** are the merger of the electromagnetic and weak nuclear forces.

The Higgs has its charms. It reconciles the oddities of the weak force and gives particles their mass without spoiling the delicate balances of the Standard Model. But it does so at the price of added complexity. Explaining it almost certainly demands an even more complete unified theory.

The Least You Need to Know

- The Standard Model describes the world as particles of matter interacting via particles of force.
- Particles of matter fall into two basic categories, quarks and leptons.
- The model describes three of the forces of nature: electromagnetism, the weak nuclear force, and the strong nuclear force.
- Particles can be either "real" or "virtual" depending on whether they are directly observable.
- Electromagnetism and the weak nuclear force are two sides of a single electroweak force.

Chapter 6

The World of the Small

In This Chapter

- Size matters
- The forces of nature vary in strength
- The fingerprints of unification
- The Planck scale

One of the central themes of particle physics is the relationship of energy and size. In fact, particle physics is commonly called "high-energy physics" because small size corresponds to high energy. In this chapter, we explore various aspects of this relationship, which gets to the heart of how nature is put together.

Small Is Different

When I was a kid, one of my favorite books was *Danny Dunn and the Smallifying Machine*, which tells the story of a teenager and his friends who carelessly stumble into a machine that shrinks them to insect size. Just getting something to drink becomes an adventure for them. They find some rainwater in a teacup but have to hit it with a nail (which, to them, is a sledgehammer) to break the surface tension of the water—an effect that people of normal size seldom notice but that looms large when one is shrunk down. The book has its scientific flaws, but the basic point is sound: the microscopic world is not just a scaled-down version of the macroscopic one.

The movie *The Incredible Shrinking Man* explores the same theme. At the end, the miniaturized protagonist reconciles himself to the prospect of continuing to shrink and encounter new adventures. If the film hadn't stopped there, skirmishes with cats and spiders would have given way to surreal scenes of befriending parameciums, falling between the cracks of molecules, and swatting away electrons. The whole fabric of nature would have changed around him as new particles and new laws of physics came into play. It would have become quite a challenge for the Hollywood special-effects department of 1957.

The driver of these changing conditions is energy. In proportion to their size, small things lead more energetic lives and endure greater fluctuations in energy than we larger things do. It's a raucous place down there. Physicists who want to visit must take heed.

The first connection between energy and size is that seeing small things requires high-energy photons. To take a picture of something tiny, we need light with a comparably short wavelength. If the wavelength is too long, the objects we're studying will slip in between wave crests and diffuse the light waves, producing a featureless blur. The wavelength sets the smallest possible pixel size in an image. It also determines the amount of energy carried by each photon of light: the shorter the wavelength, the more energetic the photons.

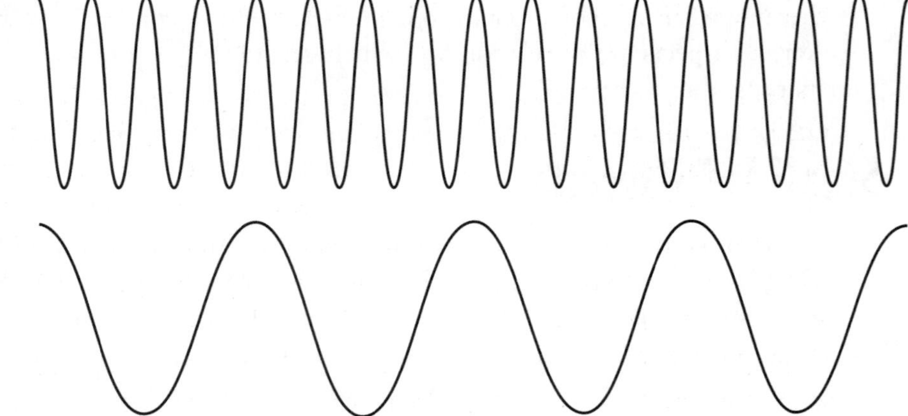

Particle wavelength and energy are two ends of a see-saw; when one goes down, the other goes up.

Compared to violet light, red light has a wavelength twice as long, and each red photon carries half as much energy. Infrared and radio photons are even longer and weaker, and ultraviolet and x-ray ones are shorter and more potent. The amount of energy in an individual photon is minuscule, but the mere fact that it varies with wavelength has consequences we can see for ourselves. For instance, to view the heat given off by a person, we need infrared goggles, whereas hotter objects give off a visible glow. As we raise their temperature, they're red hot, then orange hot, then bluish hot. The increasing energy (represented by temperature) means decreasing wavelength (represented by color).

Quantum Leap

Particle physicists and string theorists typically measure energy in units of electron-volts, abbreviated eV. Because energy is related to size (by quantum principles), to mass (by $E = mc^2$), and to temperature (by the laws of heat), physicists use electron-volts to measure just about everything. I won't bore you with the technical definition. It's better just to think of an electron-volt as the amount of energy contained in a single photon of visible light or involved in the chemical reaction of a single molecule. A proton, when annihilated and converted to pure energy, yields just under one billion electron-volts or one giga-electron-volt (GeV).

Getting a Grip on Particles

Physicists don't just passively observe particles. They toy with them, pair them up, and bust them open so they can figure out how they tick. It takes finesse to manipulate such small objects, and although we normally think of finesse and brute strength as opposites, they shade into each other in the microworld.

The reason for this is that particles aren't sitting around passively waiting for us to manipulate them. To the contrary, they're rather stubborn. Herding them is a bit like herding cats. Forget about cats in the plural; just try to herd a single cat. It's a process of negotiation. If you want her to hang out in the kitchen with you, you need to let her move at her own pace; the moment you try to force the pace, she scurries off, and you lose track of where she is. To lure her back into the kitchen is probably going to take some energy on your part.

Similarly, the more control you have over a particle's position, the less you can exert over its momentum. To guide particles where you want them to go, you need to give them leeway to spread out in momentum, and that means you need to start them off with a lot of momentum—hence a lot of energy. This tradeoff is none other than the famous Heisenberg uncertainty principle.

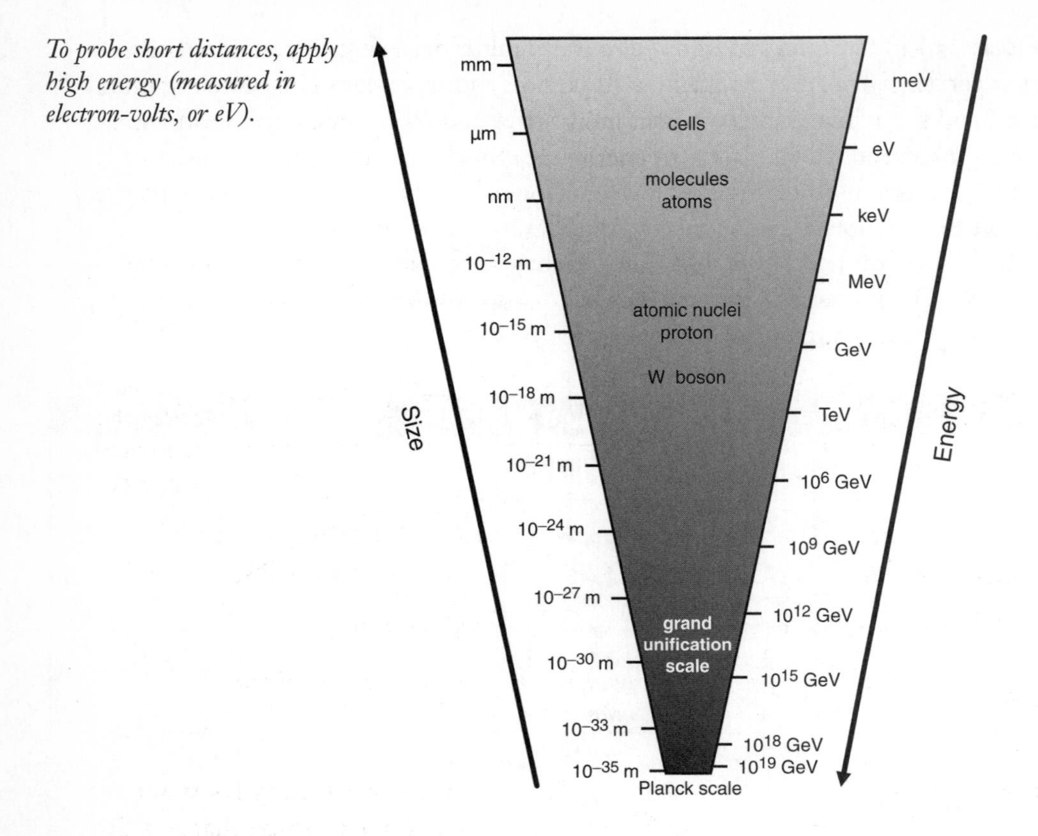

To probe short distances, apply high energy (measured in electron-volts, or eV).

An elegant way of thinking about the tradeoff is in terms of waves. Quantum theory describes not just photons but all particles as waves, indicating the range of locations they can occupy. Typically, such a wave is actually a mixture of pure waves, just as a sound wave is a mixture of pure tones. The wavelength of each pure wave corresponds to one value of momentum. By varying the number of pure waves added together in the mix, we make a tradeoff between control over position and control over momentum.

For example, if the particle has a specific momentum, the wave describing it is a single pure wave with a specific wavelength. Such a wave has neither beginning nor end. The particle could literally be located anywhere in the universe. The perfect control over momentum means utter lack of control over position.

At the other extreme, if a particle has a specific position, the wave must be a mixture of an infinite number of pure waves, arranged to cancel one another out except at that single location. Those waves correspond to an infinite range of momentum. The perfect control over position means utter lack of control over momentum.

Usually a particle falls somewhere in between these extremes. It lingers in a limited range of positions and has a limited range of momentum values. The more confined in position it is, the more spread out in momentum, and vice versa. The momentum spreads to high values, entailing high energy.

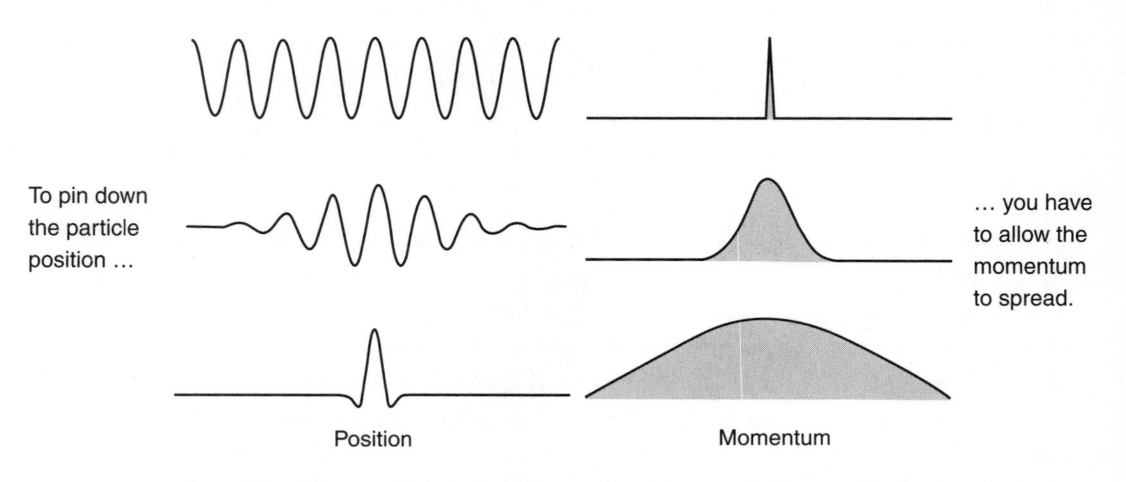

A particle with a specific momentum sprawls out in space, and a particle that is confined in space has a range of momentum values.

So the price of precision tracking is energy. That's also true when particles aren't freely moving. If you pick one up with a pair of tweezers (or something that acts like tweezers, like an electric field), the wave describing the particle has to buckle to fit in. The gap formed by the tweezers sets the maximum wavelength the particle can have and, therefore, its minimum energy. As you pinch the particle, it gets more energetic, and you have to be the one to supply that energy.

With enough energy, you can keep a firm enough grip on two particles to bring them close together or force them to collide. Particles also get whipped to high speed in extreme settings, such as explosions and the cores of stars. When their momentum is high, particles can be very tightly circumscribed. Any encounters they have with other particles occur at very short distances, so that their fine details become evident. That's why extreme situations require physicists to apply theories of the smallest constituents of matter.

Particles Come Out

Like a boisterous wedding reception that gets even your dad to do the YMCA dance, the energy of the microworld can bring out the most wallflower of particles. New

def•i•ni•tion

The **Compton wavelength** is the wavelength for which a particle has enough energy to create a whole new copy of itself.

particles can materialize if the energy crackling in the air is equivalent to their mass, according to Einstein's equation $E = mc^2$. The distance corresponding to this energy threshold is called the *Compton wavelength*. Every particle has its own value of the Compton wavelength, depending on its mass. For the electron, it's about 2.5 trillionths of a meter or 2.5 picometers. The energy it takes to guide two particles within 2.5 picometers of each other is also enough to create brand-new electrons.

This phenomenon means that new forces can kick in at short distances. An example is the weak nuclear force. Suppose two particles are brought within 10^{-17} meter of each other. That's the Compton wavelength of the W boson, which carries the weak force. The approaching particles have enough energy to create W bosons and therefore to exert the weak force on each other. If the distance between them is larger than 10^{-17} meter, there's too little energy to create these bosons, and the weak force barely operates, which is how it came to be known as "weak." If the distance is smaller than 10^{-17} meter, the energy rises still higher, and things far more exotic than the W boson might come out of the woodwork.

The two particles are like kids playing a game of catch. The closer they are, the greater the variety of balls they can throw. If they're 60 feet apart, they can throw a baseball back and forth, like particles swapping photons. If they're 25 feet apart, they can choose between the baseball and a dodgeball, like particles able to exchange either photons or W bosons. At 15 feet, they can pass a basketball, too; the basketball represents a force that physicists have yet to discover because they haven't ever gotten particles to come that close together.

Particle Groupies

The electron we know and love isn't really an electron. The iconic particle of the computer age, the basis of all things electronic, is like a rock star or movie star we never get to see for all the handlers, paparazzi, and autograph-seekers that surround it. The true electron is a pinpoint nugget of mass and electric charge surrounded by an entourage of so-called virtual particles—short-lived groupies that can't be seen individually, but collectively screen the genuine article from view. As we will see, this entourage hides not only the particle but also the essential unity of physics.

Virtual particles fill space even when it appears to be completely empty; they're a consequence of the irrepressible bounciness of quantum fields (see Chapter 5). If we throw a real electron—that is, a long-lived one—in among them, the virtual particles rearrange themselves into an entourage. Some of the virtual particles are negatively charged; others are positively charged. The real electron, having a negative charge itself, repels the negatively charged virtual particles and attracts the positively charged ones.

Let's think of pairs of virtual particles as boyfriend-girlfriend couples in a throng of groupies around a celebrity. The star is something of a ladies' man, so the girlfriends take an interest. The boyfriends, apart from feeling jealous, are put off by his liking for pedicures and expensive fashions. By pushing away the guys, the star reduces the testosterone charge in his vicinity. Similarly, the electron reduces the density of negative charge in its vicinity, partly offsetting its own charge. Using high energy, which lets us probe short distances, we can muscle our way through the crowd and start to see the electron for what it really is. It's actually more impressive in the flesh than it appears from a distance.

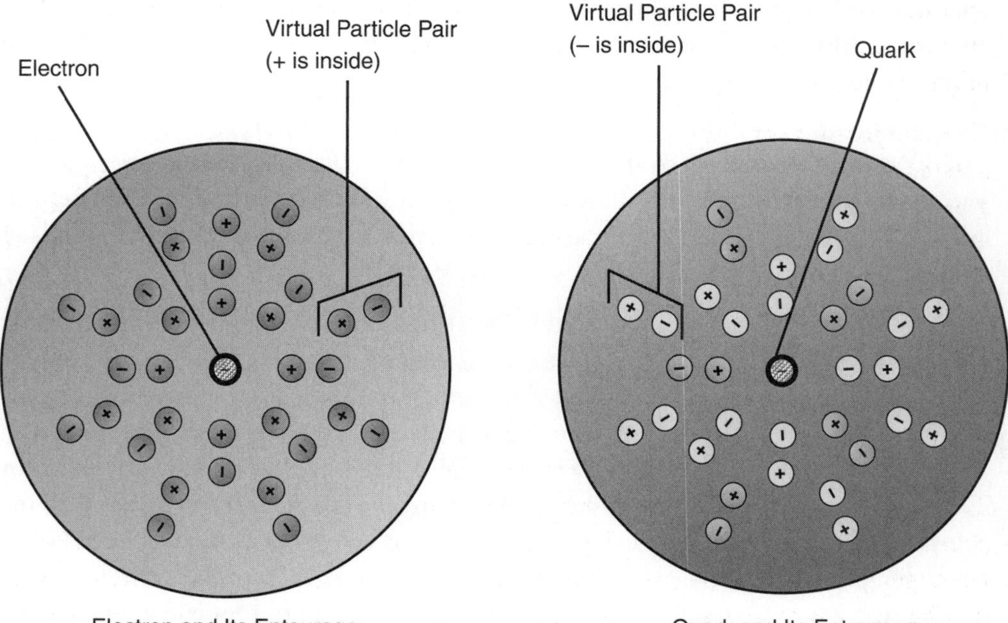

Virtual particles surround electrons and quarks and either offset or amplify their charge.

Now let's consider a quark. Its electric charge gets offset in the same way an electron's does, but it also has another type of charge: color charge. The color charge attracts a whole other set of groupies, most of which align exactly the opposite way. In terms of the metaphor, the celebrity may have an additional distinguishing characteristic besides gender. Maybe he's a great soccer midfielder. So a male groupie faces a quandary. Is he repelled by the pedicures or attracted by the playmaking? When push comes to shove, men care more about sports than about looks. The celebrity attracts male fans and increases the density of guys in his vicinity.

For the quark, the result is that the groupies amplify rather than diminish the color charge. In contrast to the electron, getting close to the quark is a letdown. It cuts a sorry figure on its own. Much of the nuclear force it appears to exert is actually generated by its entourage.

Virtual particles sound like an unreal concept, but they have a very real effect on the observed strength of the forces of nature. If we put an electron into a particle accelerator and crank up the energy, we penetrate close to the center of the entourage, the nugget becomes more conspicuous, and the total charge appears to increase. Consequently, the strength of the electromagnetic force exerted by the electron increases—by as much as 10 percent over the range of energies probed by today's particle accelerators.

If we do the same with a quark, again the entourage becomes steadily less important and the nugget more so. In this case, however, the entourage acts to amplify the quark's own color charge, so the total color charge—and therefore the strength of the strong nuclear force—decreases as we move inward. Over the range of energies probed by today's particle accelerators, the force strength varies by several hundred percent.

def•i•ni•tion

The **grand unification scale** is the distance (or, equivalently, energy) at which electromagnetism, the weak nuclear force, and the strong nuclear force become equal in strength.

In this figure, notice what's happening. As the energy goes up, the size scale shrinks, the electromagnetic force strengthens, and the strong force weakens. The weak nuclear force, for its part, also weakens. Eventually, all three forces become nearly equal in strength. It happens at a distance of about 10^{-31} meter, which is known as the *grand unification scale*. Needless to say, that's pretty small. To probe such distances would require a particle of proportionately gargantuan energy—equivalent to the mass of 10 trillion of

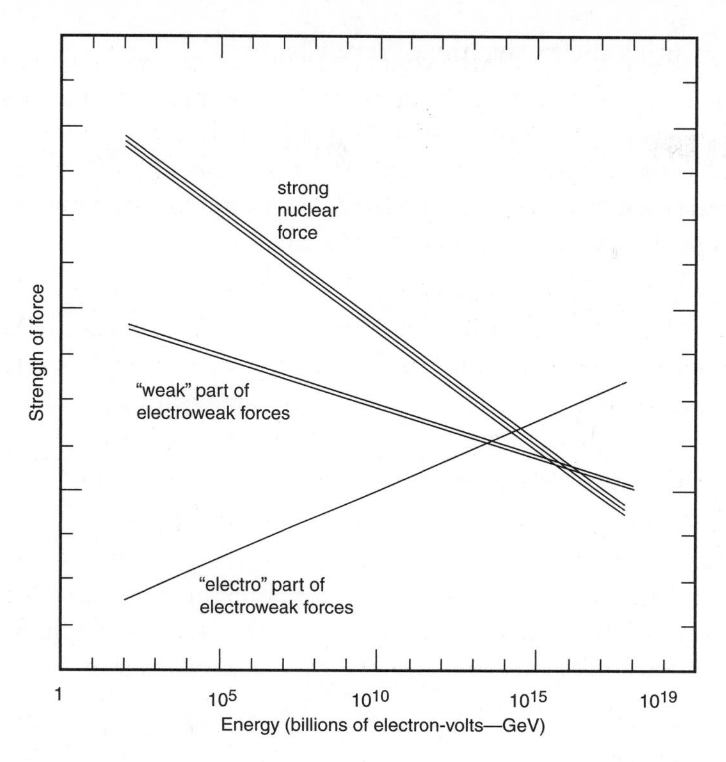

Electromagnetism intensifies with energy, while the strong and weak nuclear forces diminish, so they (nearly) meet up.

(Courtesy of Keith Dienes)

the heaviest known particles.

All Tangled Up

Why is the grand unification distance so small? Actually, the real question isn't why this scale is so small, but why *we* are so big. Our size is set by the size of the proton, which reflects a balance between electrical and strong nuclear forces. These forces are nearly scale-invariant; they vary only mildly in strength with size scale. It takes a huge ratio of sizes for their strengths to change by enough to come into balance. The proton stabilizes at a radius of 10^{-15} meter.

The mere fact that the forces become nearly equal in strength is a strong sign that they're related to one another. Two lines always meet at a point, but for three to do so would be quite a coincidence. It seems that those particle entourages have been hiding nature's true unity from us. By pumping up the energy, physicists can chase the groupies away and see the three forces for what they really are: different aspects of a single force. At the same time, the fact the three lines don't *exactly* meet suggests that something is still missing. Is string theory that thing? (See Chapter 17.)

Two Types of Forces

As significant as the variation in force strength is, it's pretty mild in the grand scheme of things. If we could boost the energy of a quark by a factor of one quadrillion, we'd sap the strong force by a factor of only 40. The other forces vary even less. If we plot their strength on a graph, the variation doesn't show up unless we exaggerate the scale.

This near-constancy is a defining characteristic of electromagnetism and the two nuclear forces. These forces behave in the same way no matter what the energy or size scale is. They are fundamentally scale-invariant or, in the jargon, "renormalizable." Electromagnetism, for instance, always involves two charged particles and one photon at a time. This template stays the same no matter how extreme the conditions get. We don't ever have four charged particles converging.

Many other processes in nature are nearly scale-invariant, too. The laws of fluid flow are a classic example. They don't care about absolute size. If someone showed you pictures of a great river, a modest brook, and the rivulets in your driveway, you might not be able to tell which is which. To be sure, they're not precisely the same. A miniaturized human being can't easily take a sip of water, and fluids of different composition flow somewhat differently. But these variations can be captured by tinkering with a few quantities in the equations rather than rewriting the equations themselves. The equations fail only when we probe right down to the molecular level.

Similarly, although electromagnetism and the nuclear forces vary in strength, these variations are captured by tinkering with a few quantities such as the charge of the electron. The equations themselves stay fixed. New forces of nature may kick in at short distances, but the equations work right up to the point where that happens, and the large-scale behavior of particles is independent of these fine-scale details.

Forces can be scale-dependent, too. For these forces, absolute size matters because their strength varies dramatically with energy. Gravity is the prime example. Newton's law of gravitation indicates that the force between two bodies depends on their masses. If we double the mass of each, we intensify gravity fourfold. Likewise, if we boost the energy of two particles by a factor of one quadrillion (10^{15}), we strengthen their gravity by a factor of one quadrillion squared (10^{30}). The variation of the strong force is a pittance by comparison. Gravity and other scale-dependent forces may be nonplayers in low-energy particle reactions, but hold their own when conditions are

All Tangled Up

If gravity is a nonplayer under everyday conditions, then why does it hurt so much when I fall down? The reason is that gravity, though weak, is cumulative. Each particle in the earth exerts an undetectably small force, but there are an awful lot of particles, and their gravitation adds up. Electrical forces, by contrast, sometimes add up (for like charges) and sometimes cancel out (for unlike charges).

harsher, as they were in the early universe.

Scale-dependent forces are much harder to get a handle on. Once we know how a scale-invariant force works at one scale, we know it at all scales. But a scale-dependent force is a shape-shifting beast, ever changing in behavior. Although scale-dependent forces complicate matters, physicists think that the apparent complexity is an artifact of our imperfect understanding, and that once we apprehend the full structure, we will find it to be simpler.

The Basement of Reality

If gravity is ordinarily weak, but gets stronger so rapidly, it eventually becomes the equal of the other forces of nature. This occurs at the so-called *Planck scale*. The Planck scale marks the realm of quantum gravity, where gravity becomes a player in particle reactions and where quantum effects intrude into the behavior of gravity. The Planck scale will be a star actor throughout the rest of the book.

In size terms, the Planck scale is 10^{-35} meter. To probe it would require pumping a truly epic amount of energy—equivalent to the mass of 100 quadrillion of the heaviest known particles—into a single particle. If we tried to pump in any more, the particle would turn into a black hole, and that would be the end of the line. A black hole is almost featureless. We can't see inside it, and if we tried to probe shorter distances by cranking up the energy still further, all we'd do is create a bigger black hole. Consequently, Planck-scale objects are the smallest possible pixels in any image. Most physicists think

def•i•ni•tion

The **Planck scale** is the defining scale of quantum gravity. It can be thought of as either a very short length (the Planck length) or very high energy (the Planck energy).

that the Planck scale is the shortest distance that has any meaning at all.

Tantalizingly, the Planck scale isn't too far from the grand unification scale I talked about previously. In other words, not only do electromagnetism and the two nuclear forces link up under extreme conditions, but gravity joins in around the same point. This convergence is a further sign they're all related. In fact, unifying some of the forces might require us to unify them all. We might call it the Three Musketeers principle: one for all and all for one.

One caveat here is that the value of the Planck scale assumes the forces continue to strengthen at the same rate no matter how high the energy gets. But it's quite possible that the trend changes, in which case the Planck scale may not be quite that small. Whatever its value, this is it—the basement of reality, the last stop on the elevator. Fathoming what happens down there is the job of string theory, loop quantum gravity, or a similar theory (see Chapter 16).

The Hierarchy of Nature

As the incredible shrinking man shrinks, he scrambles down the tree of physics. The different levels of the tree correspond to different scales of size and energy. One of the most remarkable facts about the world is that these levels remain distinct, rather than shade into an undifferentiated blur. The laws governing the world change as you descend and each set of laws works almost independently of what lies beneath it.

An economist can predict broad business trends without having to understand consumers' detailed buying habits. A doctor can suture wounds and fill out insurance paperwork without knowing what makes up the atoms in your body. New properties emerge at each level. Atoms don't breathe; individual consumers don't go into recession. These are collective attributes of large numbers of atoms or people. The principles that govern organization are at least as important as the make-up of the individual building blocks.

The same goes for the Standard Model. It's a description not only of the constituents of nature but also of how constituents fit together in a hierarchy. The hierarchy isolates macroscopic creatures such as us from what goes on at the very finest scales. The scale-invariant forces work much the same no matter what the size is, so they bear little imprint of the deep structure, and the scale-dependent forces wither under the mild conditions of everyday existence. At the same time, the Standard Model offers enough hints of a deep structure to indicate that nature is not an infinite hierarchy but has a lowest level. The yawning gap between us and the Planck scale is what makes the search for a quantum theory of gravity a journey into the unknown.

The Least You Need to Know

- Small scales correspond to high energies.
- At a very short distance, known as the grand unification scale, the three quantum forces become nearly equal in strength.
- At a still smaller distance, known as the Planck scale, gravity seems to join in, too.
- Gravity, unlike the three quantum forces, is not scale-invariant, which makes it hard to understand.

Part 3 The Need for Unity

Although Einstein's theories and quantum theory are incredibly powerful, they have their kryptonite. They can predict the behavior of atoms and planets with exquisite precision but crumple into a writhing heap when confronted with questions a five-year-old might ask. What's interesting is that where one theory does poorly, the other tends to do well—suggesting that unifying them would solve their problems. Unification becomes absolutely essential in extreme settings, such as black holes and the big bang, which neither theory can handle on its own.

Chapter 7

Why Unify?

In This Chapter

- Reasons to unify the theories
- The problem of frozen time
- Where the Standard Model fails
- The bane of the cosmological constant
- Why haven't past efforts to unify worked?

On a purely practical level, there's no real need to unify the general theory of relativity and quantum theory. They're both incredibly successful experimentally and conceptually. But on an intellectual level, each is missing something. In a case of yin and yang, quantum theory may fill the holes in relativity and relativity may plug quantum theory's gaps.

Theories of the World, Unite!

The general theory of relativity was not even a year old and quantum theory was still in its birth throes when Albert Einstein recognized that the two theories would have to be reconciled. Their worldviews are compelling yet incompatible, and the clash comes out in three ways.

- **Extreme situations.** Ordinarily, when gravity is strong, quantum effects are weak, and when quantum effects are strong, gravity isn't. But when both are in full force, they feed off each other, and only a unified theory can tell what will happen. Situations where this happens include black holes (see Chapter 8), the big bang (see Chapter 9), and would-be time machines (see Chapter 10).
- **Conceptual footings.** Nature fits together seamlessly, yet the two theories don't. Quantum theory treats space and time as fixed and absolute—which general relativity denies. General relativity treats objects as having definite properties, such as position and velocity—which quantum theory denies. The depth of their incompatibility becomes evident when physicists try to describe gravity in quantum terms (see Chapter 11).
- **Loose ends.** Each theory has shortcomings that require some deeper theory to fix. Relativity is riddled with holes; quantum theory is a hall of mirrors; and the Standard Model of particles passes the buck on crucial questions. These issues are the subject of this chapter.

General Relativity vs. Quantum Theory

Aspect	Relativity	Quantum
Basic idea	Space and time are unified and behave like a big sheet of rubber	Matter and energy are divided into chunks
What it explains	Gravity	Electricity, magnetism, nuclear forces
How it explains them	Matter distorts spacetime, and spacetime guides matter	Particles interact
Poster child	Black hole	Mighty atom
Sample use	Orbits of celestial bodies	Chemical reactions
View of spacetime	Dynamic	Static
Properties of matter	Definite	Probabilistic

Aspect	Relativity	Quantum
Toy that represents it	Silly Putty	Legos
Worst failing	Predicts that matter reaches infinite density in black holes	Nobody knows what the theory really means

Woe with Einstein

The next time you're running late for an appointment, use this perfect excuse: physicists think there may be no such thing as time, in which case there's no such thing as lateness, either. The future does not come to pass. It already exists, so you are on time no matter what you do. Try it; I'm sure it'll work.

Considering how obsessed most people are about time—even those who profess to be laid-back get upset when it's *their* time that's being wasted—you'd at least hope that physicists and philosophers of physics could figure out what time is. Sorry. Einstein's theories describe what space and time do, but not what they are. Are they things in and of themselves, with a reality independent of stars, galaxies, and their other contents? Or are they merely artificial devices to describe how objects are related to one another?

The experts argue back and forth and have invented these scenarios to help probe the issue.

In the Loop

Are they [space and time] like a canvas onto which an artist paints; they exist whether or not the artist paints on them? Or are they akin to parenthood; there is no parenthood until there are parents and children?

—John Norton, University of Pittsburgh

That Empty Feeling

Imagine a patch of spacetime that's totally empty: no stars, no galaxies, just void. Outside this patch, the equations of relativity say what the shape of spacetime is; the distribution of matter determines it. Inside, however, relativity is ambiguous. There, spacetime can bend itself into any number of shapes as long as it dovetails with spacetime outside the patch.

So spacetime behaves like a big tent. The tent poles, which represent matter, force the fabric to assume a certain shape. But if we leave out a pole, creating the equivalent of an empty patch, part of the tent can sag or bow out or ripple unpredictably in the wind. What happens in the empty patch is random. We don't need to concoct empty patches to encounter the predicament. Everywhere we go, relativity is unable to choose among multiple possible shapes of spacetime. This predicament, known as the "hole argument," greatly puzzled Einstein.

Relativity allows a hole in spacetime to be filled in many different ways.

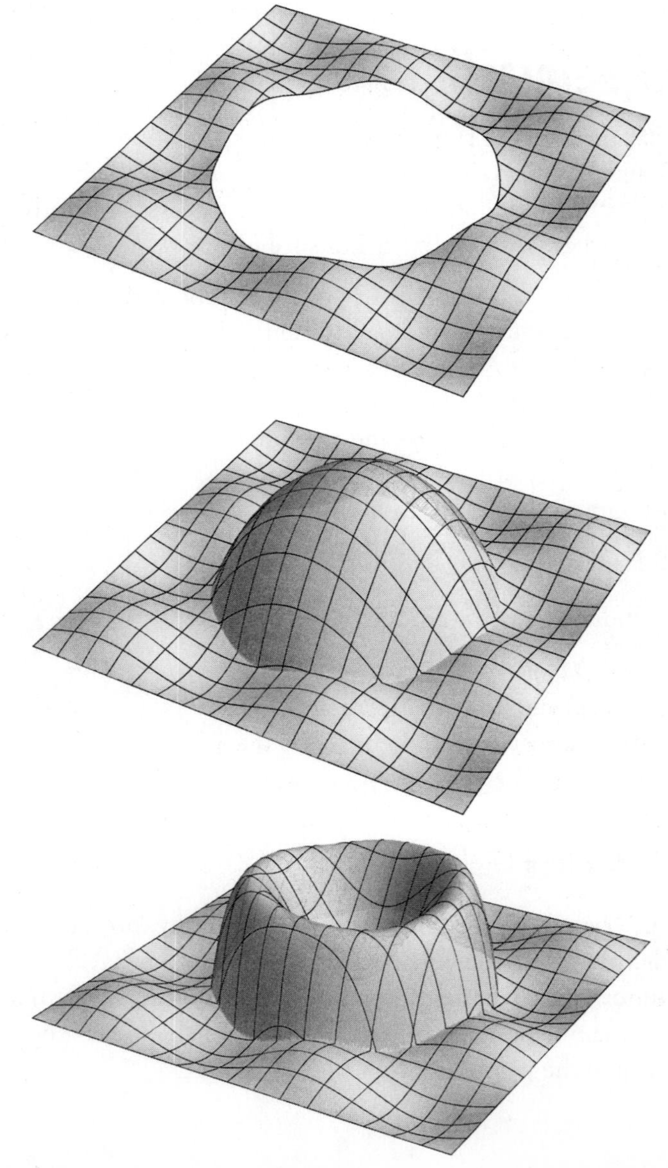

At first glance, this randomness violates the concept of determinism—whereby everything that happens, happens for a reason. For determinism to hold, those shapes must be mathematical artifacts—equivalent ways of describing the same set of relationships among objects. Spacetime might still have an existence independent of its contents, but it can't be as simple as a blank canvas that a painter fills.

Trying to Find the Time

A related conundrum arises when we break spacetime into space and time to study how space morphs over time. As matter jostles around, its gravity changes, so space should take on a new shape. But the equations indicate that its shape does not, in fact, change. The world according to relativity is locked in place, like a children's game when someone shouts "Freeze!" All measurable quantities remain constant. Physicists call this the *problem of frozen time*. Mathematically, the little *t* denoting time drops out of the equations, and the theory struggles to explain why the world around us is dynamic and ever-changing.

At the least, it suggests that mathematical time is not the same as clock time. Some scientists go to extremes and say there is no such thing as time, period. Maybe time is like money, having no meaning on its own, but acting as a convenient means of exchange. When we buy a loaf of bread, we're really bartering our labor for it. In principle, we could do away with money and undertake an intricate set of barter exchanges. Similarly, when we say that a heart beats once a second, we're really saying that it beats once per tick of a clock. In principle, we could relate heartbeats directly to the oscillations inside a clock, and to all the other processes in the universe, without ever mentioning the word "time." The world may be frozen in place, but the cat's cradle of interrelationships produce the illusion of change.

def•i•ni•tion

The **problem of frozen time,** also known simply as the "problem of time," is the conundrum that according to the general theory of relativity, the world should be static and unchanging.

In the Loop

Time is an effect of our ignorance of the details of the world. If we had complete knowledge of all the details of the world, we would not have the sensation of the flow of time.

—Carlo Rovelli, University of the Mediterranean

For most researchers, getting rid of time flies in the face of common sense. So if you ever try out your excuse that time and punctuality are illusions, be prepared for the response: "Prove it." Relativity can't offer any such proof; the philosophical dilemma has no impact on its quantitative predictions. Only by uniting it with quantum theory might physicists find a way to unravel what time really is. And if you can come up with a unified theory, I think the person you're meeting will forgive your lateness.

The Standard Model Gets Ratty

So are you asking yourself what the heck quantum theory means? That question has troubled physicists ever since there was such a thing as quantum theory. Unlike relativity, quantum theory evolved mostly through ad hoc steps, rather than a grand vision. The world described by the theory is wonderfully rich, but also inscrutable—it limits what we can learn about it. Physicists are especially perplexed by what happens during a measurement, when an element of randomness enters into the theory's predictions. Unification with general relativity may not magically clear away the haze, but by pushing quantum theory into new realms, it should at least inspire some new thinking.

Quantum Leap

Interestingly, quantum theory's problem with explaining the results of measurements is exactly the opposite of general relativity's. In relativity, it's hard to figure out what to measure. Space and time are so malleable that you can't pinpoint where and when the measurement occurs. But it's easy to explain the value you get, since the theory makes definite predictions. In quantum theory, it's easy to figure out what to measure, but hard to explain the value you get, since the theory makes only probabilistic predictions. It's an example of how the two theories have complementary strengths and weaknesses.

The pinnacle of quantum theory, the Standard Model of particles, makes such accurate predictions that it can't be wrong per se, but it can't be the final word either. Too many questions remain unanswered, such as:

- Why are there three quantum forces—electromagnetism and the weak and strong nuclear forces—and why are they progressively more elaborate?
- What happens to electromagnetism on very fine scales? The nuclear forces weaken, yet electromagnetism strengthens without limit. Some new effects must intervene to tame it.

- Why is the weak nuclear force lopsided, treating seemingly equivalent particles differently?
- Why does matter exist to begin with? The particles of force emerge organically from the principles underlying the model, but the particles of matter have to be put in manually.
- Why do particles of matter come in two categories, quarks (the nuclear particles) and leptons (such as electrons and neutrinos), and in three progressively heavier families?
- How do particles in different categories have such closely related properties, such as a common unit of electric charge?

For answers, physicists look to unification. So let's take a closer look at a couple of especially vexing types of problems.

Hierarchy Problems

A *hierarchy problem* sounds like something Dilbert would whine about. The hierarchical modern corporation is truly a marvel of organizational prowess, the marvel being that a company manages to get anything done at all. The Standard Model, too, is hierarchical, and physicists sometimes marvel that it works as well as it does. Here are three of its hierarchies and the problems they create:

- **The range of elementary particle masses.** The top dog is the top quark, which weighs much more (a million times) than the humble electron, let alone lowly neutrinos (10 trillion times more). These values are about as sensible as a typical corporate pay scale. The model accepts them as a fact of life, without explanation.
- **The range in the behavior of forces.** At a high enough energy, the forces of nature begin to unite. Electromagnetism and the weak nuclear force merge together at an energy roughly equivalent to the mass of one top quark. But to meld the combined electroweak forces with the strong nuclear force is projected to take 10 trillion times as much energy. Why do forces have this huge difference?

def•i•ni•tion

A **hierarchy problem** arises when processes occur on widely varying scales. An example is the Higgs hierarchy problem, in which the energy of the Higgs particle is much less than that of the processes that determine it.

- **The full range of energy scales in nature.** The highest we can go is the Planck scale (see Chapter 6) and the lowest is empty space. Empty really means as empty as possible. Even the best possible vacuum is still threaded by electromagnetic and other fields. Being quantum, these fields can't be zeroed out. Their irreducible energy is the same everywhere in space and in time, so physicists call it the *cosmological constant*. Observations suggest it's equivalent to a handful of particles per cubic meter. The Standard Model can't begin to explain such a low value.

def•i•ni•tion

The **cosmological constant** is the energy present in empty space.

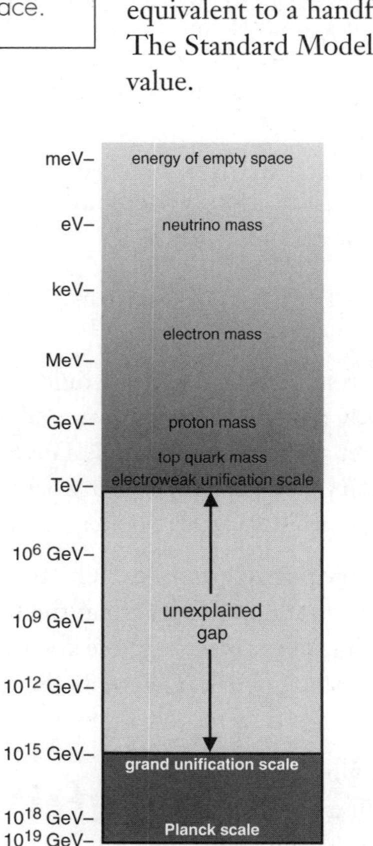

Processes in physics arise at vastly different energies for unknown reasons.

Quantum theory has so little to say about these hierarchies because it has a socialist streak; it acts to break down hierarchies and make everything equal. The most natural value for quantities such as mass is either zero or extremely high, such as the grand unification scale or the Planck scale. If a value lies somewhere in the middle, we have to jump through theoretical hoops to justify it.

In the case of the particle masses, their most natural value is zero. To explain their actual values, physicists had to build a special process into Standard Model. The process, involving the so-called Higgs particle, is widely considered ad hoc. No disrespect to Dr. Higgs, but if physicists were voting for their least favorite particle, the Higgs would probably win. One of the model's own fathers called it the "toilet" of the Standard Model.

In the case of the force strengths, the Higgs is again the troublemaker. Its problem, as with certain people, is excessive self-love. The Higgs interacts with all particles, including itself. This self-interaction generates extra energy that acts as extra mass. Most other particles with a narcissistic tendency have counterbalancing traits that put a cap on the extra energy—but not the Higgs. Its mass should get pumped all the way up to the grand unification scale, carrying the merger of electromagnetism and the weak nuclear force with it.

The Standard Model maintains the hierarchy by fiat—by endowing the Higgs particle with static properties that cancel its hierarchy-busting tendencies. The cancellation has to be good to 14 decimal places. That strikes physicists as ad hoc. They have a hunch that some hidden dynamic process limits the Higgs mass.

In the case of the cosmological constant, the most natural value is again the grand unification scale. Adding up all the field fluctuations implies an energy equivalent to more than a googol (10^{100}) top quarks per cubic meter. In other words, empty space should be chock-full of energy. Adding anything to it should be like spitting in the sea.

For most situations, this energy stored in the vacuum is unimportant. It's paper money, like the value of your house. Whether it's $50,000 or a million dollars, the money locked up in your home doesn't help you pay for groceries. Similarly, for the three forces of the Standard Model, the huge vacuum energy is immaterial because it's constant everywhere, and these forces respond to energy differences. For gravity, though, the vacuum energy does matter. The strength of gravity depends on the total absolute amount of energy, counting both empty space and the particles within it.

The Standard Model postulates some static energy—a fixed amount woven into the fabric of space—to counterbalance the field fluctuations. As with the Higgs, the cancellation of static and dynamic energy has to be improbably exact. Physicists call this oddity the *cosmological constant problem*. It makes Dilbert's company look like a paragon of rationality.

def•i•ni•tion

The **cosmological constant problem** is the mystery that the cosmological constant should be huge but isn't.

A Matter of Antimatter

Don't take this personally, but you shouldn't exist. You're made of matter, quarks and leptons, as is the bulk of the visible material in the universe. The cosmos contains only a smattering of antimatter, consisting of antiquarks and antileptons, which is good news. Whenever matter and antimatter touch, they annihilate in a blaze of energy. So if the cosmos contained equal amounts of matter and antimatter, it would soon have none of either. All that would remain is an inferno of energy—a universe full of light and nothing to illuminate.

Clearly, matter and antimatter must have gotten out of balance at some point. An excess of one part in a billion would have been enough. That is, for every 1,000,000,001 quarks, there used to be 1,000,000,000 antiquarks. After they paired off and blew up, only a single particle remained. Repeated over the cosmos, this process would leave enough matter to make up all the observed galaxies.

But how did even that modest excess arise? In the Standard Model, matter and antimatter behave almost identically. Although they do differ slightly, physicists don't know how this slight difference could have generated an excess of matter. Particle reactions are two-way streets: if they can create an excess of matter, they can just as easily erase it. For the reactions to go one way and not the other, conditions must be sufficiently out of kilter, and according to the Standard Model, they never were. The answer to the puzzle of your existence must come from beyond the Standard Model.

A Punch in the GUT

Although the Standard Model brings together the three known quantum forces under one roof, it really unites only two of them: electromagnetism and the weak nuclear force. The strong nuclear force remains distinct. Many of the problems with the Standard Model stem from this incomplete unification.

def•i•ni•tion

A **grand unified theory (GUT)** is a theory that describes electromagnetism, the weak nuclear force, and the strong nuclear force as aspects of a single undifferentiated force. Physicists have proposed many such theories, but none yet works.

To get all three forces to tie the knot, physicists since the early 1970s have sought to develop a so-called *grand unified theory (GUT)*. Grand unification is not a full unification—it still leaves gravity out in the cold—but it does fuse all the known quantum forces and particles. The electroweak lies down with the strong, the quark with the lepton, electric charge with quark color. A GUT would explain why diverse aspects of nature have so much in common.

Among the many arguments for a GUT, notice that two of the hierarchies mentioned in the previous section involve the number 10 trillion. The electroweak and strong nuclear forces become equal in strength at an energy equivalent to the mass of 10 trillion top quarks. Top quarks are 10 trillion times as massive as neutrinos. That's an intriguing coincidence. It suggests that a GUT may explain the anomalously low mass of neutrinos, which makes no sense in the Standard Model anyway.

There are different ways to formulate a GUT. All introduce a new type of particle, dubbed the "X" particle, which carries a new type of force that converts a quark to a lepton or vice versa. This particle is a mixed blessing. It might have helped rig the universe in favor of matter, but it would also destabilize protons, the bedrock of atomic nuclei. If the quarks inside protons could transform into leptons, they eventually would, and the proton would fall apart. It's a rare process, but it should be observable, and physicists have yet to see it happen. That null result rules out the first round of proposed GUTs.

What's tantalizing is that these GUTs *almost* work. They're so close. It's as though they're missing just one more concept to round them out. For that concept, physicists need to look to a more ambitious unification—one that also connects to the behavior of gravity and spacetime. It's another example of how general relativity and quantum theory may be the answer to each other's needs.

The Least You Need to Know

- Relativity theory has a number of loose ends, not least the question of what time really is.
- Quantum theory begs for a firm conceptual grounding.
- The Standard Model of particles explains particles, but nothing explains the model.
- A partial unification, called grand unification, doesn't work, and no one knows why. The answers may lie in the full unification of quantum theory with relativity.

Chapter 8

Black Holes

In This Chapter

- The definition of a black hole
- Anatomy of a black hole
- Stephen Hawking's great discovery
- Do black holes erase information?

Almost everything about black holes cries out for a unified theory of physics. At the center of one, gravity is so intense that general relativity breaks down. At the perimeter, particle pairs are torn asunder and quantum theory chokes. No wonder black holes are everyone's top choice for the most bizarre objects in the universe!

Down the Drain

Not to alarm you, but there's a black hole of sorts in your bathtub. It shouldn't do you much harm unless you're a small bug. The way water drains out of a tub is just what happens to matter at a black hole out in the cosmos. At the edges of the tub, water flows languidly, but as it approaches the drain, it get funneled inward and picks up speed. Any paramecium that fails to keep a safe distance from the drain passes a point of no return.

Closer to the drain, its distress calls get swept down, too, because waves can't buck the flow either. In an emptying tub, no one can hear a paramecium scream.

A black hole is a kind of cosmic drain where material is drawn in, not by the flow of water but by the force of gravity. Far from the hole, the gravitation is not especially strong, but as anything approaches it, gravity intensifies. Any rocket that fails to keep a safe distance passes a point of no return where it would have to accelerate to the speed of light to break free from the hole's gravitational pull. And the astronauts on board can't send out a distress signal because light can't escape either.

All Tangled Up

Do black holes suck up everything like an all-powerful vacuum cleaner? As bad as black holes are, they aren't *that* bad. Their gravity is irresistible only if something ventures too close. From a distance, they are no more powerful than any other object. If our sun turned into a black hole, heaven forbid, it would get awfully dark here on Earth, but our planet would continue orbiting. Even if you wanted to fall into a black hole, it's not as easy as you'd think. It presents a small target, and unless you aimed straight for it, you'd fly right past.

But it gets worse. At least the paramecium flows into the sewer pipe and lives happily ever after. But there is no sewer to carry away the matter that falls into a black hole. Whatever enters has to live within its minuscule confines. A black hole is the ultimate trash compactor, creating the tightest wad of matter possible—so tight that it can't even be called matter anymore. According to general relativity, the poor astronauts get completely annihilated and converted into pure gravitation, making the black hole that much bigger, stronger, and deadlier for the next ship that passes by.

Quantum Leap

If nothing can escape a black hole, how does gravity get out? According to general relativity, it doesn't need to. Gravity is determined by the shape of spacetime, which is established when the hole first forms. Once it's set, it's set—there's no need for a force to continue traveling through space. Relativity says that only when gravity *changes* does something need to propagate outward. That "something" is a gravitational wave (see Chapter 11).

According to Newton's law of gravity, a black hole isn't possible; a powerful rocket, firing for long enough or navigating a circuitous path, can avoid the clutches of even the most massive planet or star. But in general relativity, when enough matter gets packed into a small volume, its gravity goes wacko, feeding on itself. There's no escaping then.

How to Make a Black Hole—or Not

For relativity theory, the real question isn't why something would turn into a black hole but why everything doesn't turn into one. Fortunately for us, various effects ride to the rescue. Earth is saved by the integrity of its materials; the electrical repulsion between atoms in rocks resists crunching. The sun is saved by its nuclear reactions; they generate gas pressure that counterbalances gravity. The galaxy as a whole is saved by the momentum of the sun and other stars; they buzz around too fast to fall to the center and get crunched together.

Quantum Leap

A black hole with the mass of the sun is about 3 kilometers in radius (smaller if it spins). So to turn the sun into a black hole, we'd have to squeeze it from its current radius of about 700,000 kilometers (100 times bigger than Earth) to the size of a small town. The radius of a black hole scales up with mass. One that's a million times heavier than the sun has a radius of 3 million kilometers.

Stars become vulnerable when they run out of fuel and can no longer generate energy to resist gravity. Their constituent particles make a last stand because particles of matter naturally resist being shoved too close together (see Chapter 4). In the sun's case, they'll succeed in holding out. For gravity to break through, what remains of a star at the end of its life must be at least three times more massive than the sun. For dead stars of this size or greater, no known effect can keep gravity at bay. It finally wins.

Types of Black Holes

Black holes come in three basic types:

- **Stellar black holes.** These form when a large star runs out of nuclear fuel and collapses under its own weight.
- **Supermassive black holes.** Found in the cores of galaxies, they have the mass of millions or billions of suns. Astronomers don't quite know how they form, but one possibility is that smaller black holes merge together.

- **Mini black holes.** These have the mass of an asteroid or less. None has ever been detected, but physicists speculate they could form under the extreme conditions of the big bang or in particle collisions.

Although black holes are truly black and almost impossible to see directly, the behavior of matter near a hole gives it away. For example, stars at the center of our galaxy whip around at such a speed that they must be pulled by a mass equal to three million suns, packed into a volume smaller than our solar system. No conceivable type of star or group of stars measures up, so the culprit must be a black hole.

Some centers of galaxies are as bright, on their own, as entire galaxies. No ordinary power source, such as nuclear energy, can explain that. But a black hole can. As things fall in, they sideswipe one another, heat up, and glow brightly. The bottom line is that black holes are so common that astronomers have grown rather blasé about them. Another black hole—yawn. But for theorists trying to explain them, black holes continue to pose a formidable challenge.

Black Hole Geography

A black hole has two main parts:

- **Singularity.** This is the pinprick at the very center of the hole, where infalling matter smushes together.
- **Event horizon.** This is the perimeter of the hole, the point of no return. Once an object crosses through, it can't avoid hitting the singularity.

Which of these two parts gives you the bigger headache depends on whether you're falling into the hole or watching with horror from a safe distance.

Taking the Plunge

Nobody has been lucky enough to visit a black hole, but physicists know enough to guess what it's like. As long as astronauts have decent shields to protect them from the intense radiation and other infalling objects, the ride won't seem that bad to them. They'll wonder what all the fuss was about. They'll feel just like skydivers in free fall—that is to say, almost completely weightless.

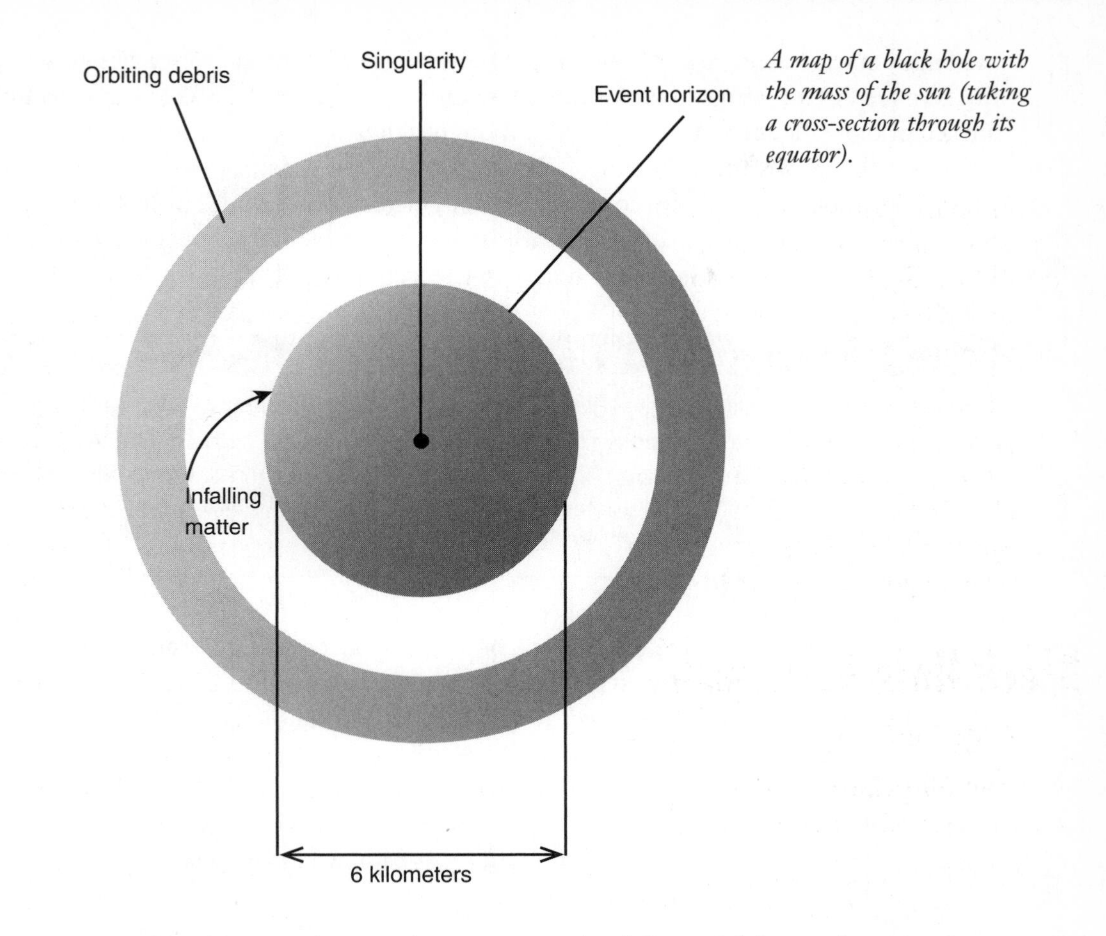

A map of a black hole with the mass of the sun (taking a cross-section through its equator).

The one thing that might give them pause is the differential force of gravity across their bodies. As with any celestial object, the gravitational pull intensifies as they get closer to the hole. If they're falling feet first, it tugs on their feet more strongly than on their heads. For small holes, this disparity would rip the astronauts apart even before they enter. For large holes, though, it would only become noticeable deep inside.

Nothing out of the ordinary happens when the astronauts cross the event horizon. Like the paramecium flowing toward the drain, they won't hit a wall or pass an ominous signpost. The point of no return is just an imaginary line in water or in space. To tell they've passed it, they'd have to plot their position against distant reference points.

Only when the astronauts reach the singularity will they say "oops" as they smack into the tightly wadded matter and get assimilated. To say they die is putting it too mildly.

Normally when we die, our molecules get recycled, providing the wherewithal for new life. We never truly vanish; we are merely scattered, and the world will always retain a trace of our existence. At a singularity, though, relativity says the astronauts literally reach the end of their timeline. Like characters on the last page of a novel, there is no "after" for them. Their molecules do not get recycled; nothing is left to carry their imprint. If the finality of their death strikes you as incomprehensible, it should. Physicists consider this a failing of relativity theory.

Quantum Leap

As you cross the event horizon, you're moving at the speed of light, so it doesn't take long to smack into the singularity. For a black hole with the sun's mass, the trip lasts about seven millionths of a second. The duration scales up with mass. For a black hole a million times heavier than the sun, the trip lasts about seven seconds once you fall in, which gives you a little time to enjoy the scenery.

To those of us on the outside, the voyage of doom appears very different. As the astronauts descend, the signals they send out have to work against the gravity of the hole to reach us. Stretched and weakened, they require longer antennas and more sensitive equipment to detect. Played back through a speaker, the astronauts' voices sound slow and slurred as if they were drunk. After a while, the astronauts start to look downright pitiful. The camera feed shows them in slow-motion, and they just keep getting slower. The equipment strains to pick up their signal until at some point it can't. The poor souls have vanished without a trace.

All Tangled Up

As the astronauts fall in, do they see the entire future of the universe go by? You might think so. If we see the astronauts in slow-motion, it stands to reason that they will see us in fast-forward. But it doesn't work like that. They see only the light emitted before they pass through the event horizon. After that point, they're moving too fast for new light beams to catch up to them.

Or, more precisely, almost without a trace. The hole is now that much heavier, spins that much faster, and, if the ship carried a net electric charge, charged up that much more. The black hole has reduced a wonderful group of human beings into the rawest of raw material: mass, spin, and charge. Whether the victim is an astronaut or a pile of ball bearings of equal weight, the end result is the same. According to a principle known as the *no-hair theorem*, the hole erases every last trace of their individuality. Physicists consider this thoroughness of destruction a major problem for quantum theory.

def•i•ni•tion

The **no-hair theorem**, an implication of general relativity, holds that a black hole is characterized fully by its mass, spin, and charge.

A Singular Problem

So the difference in perspective leads to two types of theoretical failures. The singularity is where general relativity drops the ball. The theory predicts that its density is infinite. Because the curvature of spacetime depends on density, the curvature must go infinite, too, meaning that spacetime tears open like an overloaded grocery bag. In practice, infinite quantities are mathematical fictions—a warning bell that some processes beyond the scope of relativity must come into play. Quantum theory is the only place to go looking for such a process.

The saving grace of relativity is that the event horizon hides the singularity from our view. Although the theory fails, the failure is firewalled off and doesn't affect general relativity's predictions for the rest of the universe. So the singularity is like a caged tiger. It can hurt you if you get too close, but it can't go prowling the streets, so you can sleep soundly.

Or can you? Might singularities break out of their cages? One of the leading theorists on the topic, Roger Penrose of Oxford University, assures us they cannot, an idea he calls the "cosmic censorship conjecture." But many of his fellow physicists are less sanguine. Computer simulations suggest that under certain conditions, a star can collapse to a singularity without creating an event horizon in the process. Also, spiking a black hole with the right mix of particles, like feeding spinach to Popeye, can give it the strength to break free.

A singularity without an event horizon is "naked," directly exposed to our view, and it's not a pretty sight. Whereas a black hole with an event horizon is a one-way street, a naked singularity is a two-way boulevard. Relativity has no way of predicting what it might do. Things might simply pop out of it, like rabbits out of a hat. It's a version of those pesky "space invaders" that gate-crashed pre-Einsteinian theories of physics (see Chapter 3). Relativity theory was supposed to have shot them all down, yet here they are again. By allowing things to happen for no apparent reason, naked singularities undermine a bedrock concept of science: determinism, the idea that everything that happens, happens for a reason. They add a touch of black magic to the theory, deepening its failure.

Trouble on the Horizon

At the perimeter of a black hole, the plight of the two theories is reversed. Now quantum theory is the fall guy. The one-way nature of the event horizon breaks a central tenet of quantum theory: that information can never be lost. Quantum theory may not allow us to garner much information about a system, but at least systems don't go off and lose any of the information they do have.

It's bad enough that a particle could fall into a black hole and we'd lose the information it carried. What makes it worse is that most information is stored not on particles per se, but in their interrelationships with other particles. To retrieve the data, we need to compare particles with one another, which presents a challenge when some have fallen into a black hole. It's like encrypting files on our hard drive. If we lose the password, we lose our ability to open the files. A seemingly partial loss of information can lead to a total loss.

But at least we can back up our hard drive and write down our password in a safe place. We can't back up quantum information because that would mean overwriting the information on some other particle. When a particle crosses the event horizon, the outside universe loses something irreplaceable. Without the missing particle, we can't fully decrypt the data stored on the particles it was entangled with. Part or all of the data is reduced to random mush.

Although quantum theory has a deserved reputation for introducing randomness into the world, this situation is worse. When we measure a particle, we may get a random result, yet the distinctive quantum information is never truly lost. Through the process of decoherence (see Chapter 4), it scatters into the surroundings like broken glass, and in principle we can always pick up the pieces and glue them back together. But a black hole stops us from doing that.

The Quantum Trap Door

But how bad can the information loss really be? Maybe the data is hidden but not truly lost. That happens all the time. We're always losing things and finding them again, usually in the last place we left them.

The trouble is that quantum theory predicts that black holes are temporary. Just like almost everything else in this universe, they are born, live their lives, and die. The only thing that would keep them going is a law that says they can't die. Recall the quantum injunction: everything not forbidden is compulsory. That maxim applies to black holes, too. Because they can bequeath their mass, spin, and electric charge to elementary particles and call it a day, they eventually do.

If so, then maybe the hidden information will come out. The universe will be made whole again; particles will mend their broken relationships; we'll finally be able to unlock the data on our hard drive. It might take a while—no one has yet seen a black hole decay—but the universe moves at its own pace, not ours. While this idea may resolve the quantum problem, it runs headlong into relativity's commandment that nothing shall leave a black hole. So what gives? This is the question that made Stephen Hawking's name. Reconciling the injunction and the commandment makes the information loss a true paradox.

The Hawking Effect

Black holes are unstable because as stark as the vacuum of space may be, it's still threaded by electromagnetic and other fields. Even if we bleed off all their energy, these fields still have some life left in them, just as the surface of the sea bobs gently up and down even on the quietest of days. This irrepressible bounciness creates particles, which are called "virtual" because we can't ever get a fix on one before it settles back down (see Chapters 5 and 6).

Virtual particles come and go in balanced pairs, one a particle, the other an antiparticle. That way, their net momentum as well as other particle properties is zero—as it must be because the vacuum has none to spare. In a physics version of *Brigadoon*, the two particles pop into existence, enjoy a brief sojourn, then meet up and return to the vacuum whence they came.

As in *Brigadoon*, an outsider—in this case, the event horizon of a black hole—throws off this neat arrangement. The horizon breaks the cycle of creation and destruction. If the virtual pair materializes either just outside the horizon or straddling it, the particle on the inside gets separated from its partner and can't keep their appointment to meet up and fade back into the vacuum. The particle on the outside might chivalrously fall in to join it but could also climb away from the hole for good. It then gives up the opportunity to reunite with its partner. It reaches our telescopes, earns the right to be called real, and has to live with the guilt.

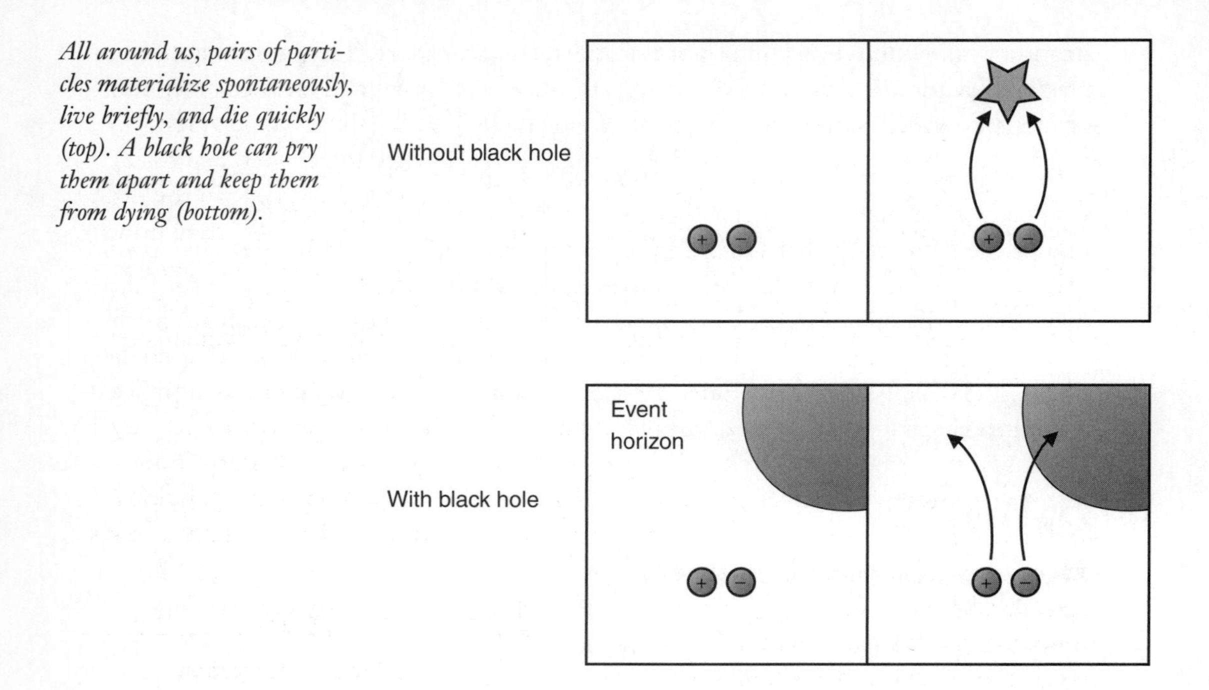

All around us, pairs of particles materialize spontaneously, live briefly, and die quickly (top). A black hole can pry them apart and keep them from dying (bottom).

Astronauts falling through the horizon would see nothing special, just the same vague quivering of virtual particles that occurs everywhere else in space. The would-be real particle has only just started to make a run for it. But those watching from a distance would see a spray of outgoing particles. For a black hole with the mass of the sun, about one particle dribbles out per second. It takes energy to power this stream, and we can't get something for nothing (even in quantum theory). There's only one place the energy can come from: the hole itself. The hole must lose mass, cannibalizing itself in order that some particles may go free.

Quantum Leap

For the Hawking effect to work, the black hole has to lose mass—to give up some of the energy it embodies, so as to propel the outgoing particles. That presents an apparent paradox. Particles falling into a black hole usually add to its mass, yet the virtual particle must somehow subtract from it. It's as if the virtual particle has *negative* energy, a concept that ordinarily is meaningless. But a black hole is no ordinary object. The whole concept of energy is connected with the shape of spacetime, and the contorted spacetime inside the horizon allows the energy of a particle to drop below zero.

This confirms the basic argument that a black hole cannot last forever. It may take trillions of years to go, or longer, but go it must. The drip-drip of particles from its perimeter depletes it and causes it to shrink down until eventually nothing is left. So what happens to any information hidden within? Do the outgoing particles spirit it away? Unfortunately, the answer is no. These particles originate very near to the event horizon—but still outside it. "Very near" isn't good enough. The particles have no way to pick up any detailed information from within. The only information they can suss out is the hole's mass, spin, and charge. The amount of energy carried by the particle reflects the mass of the hole. But whether the mass initially came from astronauts or from ball bearings, it knows not.

To sum up, a black hole swallows matter, compresses it, and then evaporates, leaving behind a mist of random particles. The origin of the matter is lost to history. This phenomenon is the famous *black-hole information paradox*. What makes it a paradox is that quantum theory has reached a conclusion (information loss) that is totally at odds with quantum theory (no information loss). If physicists had to pick the single most important puzzle for a quantum theory of gravity to solve, this might be it.

def•i•ni•tion

The **black-hole information paradox** is the conundrum that information can fall into a black hole and get destroyed, pulling the rug out from under quantum theory. It's one of the biggest unsolved problems in modern physics.

Notice that the black-hole information paradox, like the singularity problem, gets into trouble with determinism. For singularities, determinism fails because relativity cannot predict what happens to matter that reaches there. For the event horizon, determinism fails because quantum theory cannot predict what becomes of particles whose entangled partners crossed to the dark side. Not only could we not project forward in time, we could never fully reconstruct the past, how the universe got the way it is.

In the Loop

The puzzle of black hole evaporation portends a scientific revolution as sweeping as that that led to the formulation of quantum theory in the early twentieth century.

—John Preskill, Caltech

The failure of such a foundational concept cannot help but have wider implications. Black holes are the dead canary in the coalmine of physics.

Hints of Quantum Gravity

Two aspects of Hawking's analysis, in particular, suggest the need for quantum gravity. The first is that the outgoing particles begin their journey at the event horizon itself. When I described the doomed voyage of astronauts into a hole, I said that people back on Earth would eventually lose contact with the travelers as their radio waves became stretched to the point of undetectability. So how does an outgoing particle make it? Shouldn't it be stretched so much that it flatlines?

In fact, it should—unless it starts with infinite energy. And there's that word "infinite" again, a good sign that a theory is past its use-by date. Quantum gravity must come into play.

The second aspect is that the particles from the black hole look exactly like the particles being given off by a hot body, like a glowing coal or red-hot poker. Namely, they are mostly photons of light, with a range of colors as if emitted by a body with a certain temperature. For ordinary bodies, temperature and heat result from the collective motions of molecules. So does it mean a black hole consists of "molecules"? What could those "molecules" possibly be, when all the matter has been scrunched to a single point? Whatever these building blocks are, the sheer amount of heat in the hole implies that they are as small as the Planck scale—the defining distance of quantum gravity.

Black holes are the reigning example of how the conflict between general relativity and quantum theory is not just a matter of teeny-weeny doodads. A black hole millions of kilometers across sings the song of quantum gravity, too. It acts as a powerful microscope, amplifying Planck-scale effects to something we might see with our own two eyes. The same is true of cosmology, the subject of the next chapter.

The Least You Need to Know

- Black holes are a crucial test case for quantum theories of gravity.
- They show that quantum gravity applies not only to small objects but to big ones, too.
- General relativity breaks down at the singularity at the heart of the hole.
- Quantum theory fails at the event horizon at the perimeter of the hole; in particular, it predicts that information gets lost.

Chapter 9

The Big Bang

In This Chapter

- What cosmology has to do with us
- The big bang
- How current theories struggle with the bang
- What came before the beginning?
- The mystery of cosmological dark matter

Like black holes, the big bang involves extreme conditions that neither relativity nor quantum theory can explain on its own. Unlike black holes, it's something we can see with our own two eyes, if we know how to look. The fact that we exist; that it's dark at night; that outer space to our left looks about the same as outer space to our right—from these simplest of observations, we can confront the deepest mysteries of cosmology.

Roots

If you start tracing your family tree to your parents, your grandparents, your great-grandparents, and then keep on going, what happens? You follow the branches through your great-*N*-times-great-grandparents, maybe finding you do have some royal blood after all. Earlier ancestors of yours might have helped build pyramids, paint caves, or domesticate potatoes. Eventually all the family trees of all of us merge into one.

Go far enough back, and our ancestors weren't even *Homo sapiens*, but some other hominid living on the savannah. Way before that, they were furry shrews that darted left and right to avoid getting stepped on by dinosaurs. If we keep on going, we come to microbes and the roots of the tree of life. Then the lineage jumps off our planet, to the stars that created the atoms in our bodies and the cosmic primordial soup from which those stars coagulated. Ultimately, there was a time when our forebears weren't particles we'd recognize, but a purer variety. The particles dissolve into space and time, and space and time themselves dissolve into something spaceless and timeless. The universe was without form, and void.

So cosmology isn't about something way out there, something inconceivably big and old and distant. It's about *us*: who we are and how we come to be here. The broad outline of our descent is what we know as the *big bang*. The universe used to be hot, dense, and fluid. Over time, it thinned out and cooled down, which allowed bodies of increasing complexity to take shape until the magical moment 4 billion years ago when nonlife begat life on our planet.

def•i•ni•tion

The **big bang** is the evolution of the universe from a hot, dense state. Note: in this book, the term refers to the ongoing evolution. Elsewhere, many people use the term to indicate just the start of this process, the hypothetical time zero—a definition that presumes there was such a time zero, which may not be true.

Quantum Leap

To take the biology metaphor a step further, consider the old concept of ontogeny recapitulates phylogeny: the claim that the development of an embryo (ontogeny) re-enacts the evolutionary history of its species (phylogeny). A version of it holds in cosmology: ontogeny replicates ontology. The development of the universe re-enacts the laws of physics (ontology).

An amazing aspect of our cosmic family tree is that it's the same as the tree of physics (see Chapter 2). The further back in history we go, the simpler things get. In the ancient universe, matter was distributed more uniformly and consisted of fewer types of atoms. Still earlier, the distinctions among physical forces melted away. Most physicists think that everything that we now see grew from a seed made up of a single type of matter and was governed by a single type of force. Only a quantum theory of gravity can describe it.

The Meaning of the Bang

Simplicity is what makes the modern science of *cosmology* possible. In some ways, it is the simplest of all sciences. With a couple of basic processes and a few broad categories of matter and energy, we can say almost everything there is to say about the universe on very large scales.

Consider how far we can get with a couple of basic observations. First, why is the sky dark at night? It didn't have to be that way. If the universe had been around forever in its current form, starlight would have had plenty of time to seep into all corners of space, and we would see a uniform fiery glow across the sky. Wherever we looked, we'd see a star. Like trees in a forest, some would be close; some would be far; but together they'd form an unbroken wall. So the fact that we see only a sprinkling of stars means they must be a relatively recent arrival, like seedlings that have begun to reclaim a barren plot. This insight, known as Olbers' paradox after a nineteenth-century German astronomer, is a first hint of the big bang.

def•i•ni•tion

In the modern sense of the word, **cosmology** is the sub-branch of astronomy that studies the whole universe as a celestial body in its own right.

Another observation requires a telescope but is almost as simple in principle. To ascertain the overall patterns of the universe, rather than localized effects, cosmologists measure the motion of entire *galaxies*. Our solar system and all the stars we see with the naked eye belong to the Milky Way galaxy, but telescopes reveal billions of others. In the grand scheme of things, each behaves as a single unit.

We have discovered that galaxies are systematically moving away from one another. What's more, they have been fleeing each other's company for as long as anyone can tell. Deep in the past, the precursors of all the galaxies we observe were scrunched up against one another like subway riders in rush hour. Since then, they have scattered in every direction, and the universe has thinned out. Because this sounds a lot like an explosion, astronomers dubbed it the big bang.

def•i•ni•tion

A **galaxy** is a giant system of stars, gas, dust, and other material. Loosely speaking, it's the next level of organization up from the solar system.

Working backward from how fast galaxies are moving apart now and accounting for changes in their speed, all the galaxies we see would have been concentrated at a single mathematical point 13.8 billion years ago. This point bears more than a passing resemblance to the singularity at the center of a black hole (see Chapter 8). Both are places where relativity theory predicts that gravity spirals out of control and matter reaches infinite density. The main difference is that stuff falls toward the black hole singularity but shoots away from the big bang one.

This hypothetical instant of complete and total scrunching may or may not have been the dawn of time. No one knows what, if anything, occurred then. That's why I use "big bang" to mean the spreading trend rather than the moment of time zero. The big bang is not something that occurred long ago; it is an ongoing process that we are part of.

Cosmic Expansion

The motion of galaxies away from one another follows a very regular pattern—too regular, in fact, to be explained as an ordinary explosion. The farther apart galaxies are, the faster they move. Galaxies twice as far move twice as fast. Those 42 times farther apart move 42 times as fast. No matter how distant two galaxies are, their relative velocity follows this pattern. Astronomers have confirmed this trend for hugely distant galaxies by getting a fix on exploding stars inside them.

The "no matter how distant" part is the giveaway. If two galaxies are extremely widely separated, they are moving apart at the speed of light or faster. This isn't just a trick of the light giving the illusion of faster-than-light motion. If it were, cosmologists wouldn't see such a regular trend with distance.

Quantum Leap

If a pair of galaxies are 1 million light-years apart, they separate at about 20 kilometers per second. If 2 million light-years apart, they separate at about 40 kilometers per second. If 14 billion light-years apart, they separate at 300,000 kilometers per second—which is the speed of light. The farthest galaxies that astronomers have discovered are currently moving away from us at about twice the speed of light.

Brring! The Einsteinian alarm bell just went off. How can anything exceed the speed of light? Well, it can't—not if it's moving as we normally think of motion, as rocketing across space from one position to another. But Einstein's general theory of relativity allows for a second type of motion: not *through* space, but *of* space. It's the difference between an ordinary sidewalk, which you have to walk down, and a moving sidewalk, which does the work for you.

In this second type of motion, the distance between two objects increases not because the objects shift position but because the space between them gets larger. To visualize this, take a rubber band and draw some dots on it. Then stretch it. As you do, the distance between every pair of dots increases. Widely separated ones move proportionately more. From a dot's point of view, the other dots are all getting farther away. It doesn't matter which dot you pick; each sees the same thing. Galaxies in the expanding universe are like these dots.

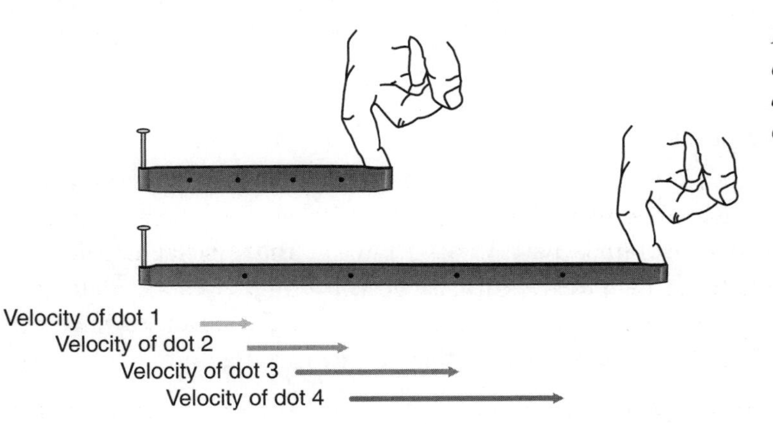

For a handy model of expanding space, draw some dots on a rubber band, pin one end, and pull.

Although this second type of motion lets us exceed the speed of light, it doesn't mean we can outrun light. After all, if space is taking us for a ride, it's taking any light beams for a ride, too. They still beat us in a one-on-one race.

The key thing to remember is not to take the term "big bang" too literally. A bomb didn't go off and hurl matter outward through space. Instead, space itself explodes. The bang did not occur at a single place, but everywhere, just as the expansion of a stretched rubber band happens everywhere along it. The difference from the rubber band is that we need to pull the band, whereas space is naturally dynamic and expands of its own accord.

All Tangled Up

What about the books and websites arguing the big bang never happened? It's true that some people do not accept that space is expanding. For discussions of the holes in these arguments, see the websites listed in Appendix B.

All Tangled Up

If space expands, why don't people, asteroids, and the solar system expand, too? The reason is that these bodies, heavenly and otherwise, have some internal cohesion. People are glued together by electromagnetic forces, the solar system by gravitational forces, and asteroids by a combination of the two. Light waves, on the other hand, do expand, which gives cosmologists a way of tracking the progress of the big bang. The waves representing material particles expand, too, which has the effect of reducing their momentum (per quantum theory), but leaves their size untouched.

The expansion of space is one of the most important discoveries of twentieth-century science. It is to cosmology what evolution is to biology: a framework for making sense of everything else.

Unwinding the Clock

By going backward in time and unwinding the effects of expansion, you can trace your family tree ever closer to quantum gravity. Way back when, the universe was denser, which implies it was hotter, which implies that particles had higher energies and behaved in steadily more bizarre ways. Some of these new effects occurred when existing ingredients came together in novel ways, others as the existing ingredients broke up. The pace of change picks up the farther back you go. Today, most astronomical processes take millions or billions of years, but long ago, events were measured in years, seconds, or fractions of a second.

- **Today** (13.8 billion years after time zero). The radius of the observable portion of the universe is now 47 billion light-years (Glyr), which reflects both how much it has grown and how far light has been able to travel since the start of the expansion.
- **Formation of solar system** (9 billion years after time zero). The sun and its planets came into being about two thirds of the way through cosmic history.
- **Earliest stars and galaxies** (100 million to 1 billion years after time zero). It took gravity this long to pack gas into the first generation of stars, galaxies, and large black holes.
- **Dark Ages** (380,000 to 100 million years after time zero). Before the first sources of light switched on, the universe was nearly pitch dark and filled with a miasma of hydrogen, helium, and lithium. None of the other chemical elements existed until stars cooked them up.

- **Formation of atoms** (380,000 to 490,000 years after time zero). Before atoms existed, electrons, nuclei, and photons made up a collective community like a kibbutz or commune. No particle had a claim on any other particle; all frolicked together. But when temperatures fell below 3,000 kelvins, electrons slowed down enough to be snared by nuclei and settle down into atoms. Photons went their own way, gradually losing energy but always retaining an imprint of the layout of material just prior to atom formation. They eventually became the *cosmic microwave background radiation* (*CMBR*), a treasure trove of data for cosmology and for fundamental physics (see Chapter 21).

def•i•ni•tion

The **cosmic microwave background radiation (CMBR),** a nearly uniform glow of microwave radiation, provides a snapshot of the universe as it was at a cosmic age of about 380,000 years. It is strong evidence of the big-bang theory as opposed to alternatives, such as the steady-state theory.

- **Formation of atomic nuclei** (1 second to 3 minutes after time zero). In the precious early moments when conditions were just right, protons and neutrons clumped to form the nuclei of the lightest elements. Most of the helium, heavy hydrogen, and heavy lithium in today's universe dates to this period. For observers probing back in time, the formation of nuclei is the end of the rope. They can glimpse this epoch by measuring the amounts of those elements, but they don't yet have any data that directly probes earlier epochs. From here on, cosmology becomes more tentative.
- **Formation of protons and neutrons** (10 microseconds after time zero). Three by three, quarks settled down into larger particles.
- **Formation of photons** (10 picoseconds, or trillionths of a second, after time zero). Around this time, electromagnetism and the weak nuclear force, which had been a single set of forces known as the electroweak forces, parted ways. Photons, the carriers of electromagnetism, emerged as distinct particles, while their ex-partners slunk away. Their dramatic breakup rippled into the brew of other particles. Before, all types of particles had the same mass (namely, zero); after, their masses ranged all over the map. As the momentous transition swept through space, it may also have rubbed out most of the antimatter in the universe, leaving behind mostly matter.

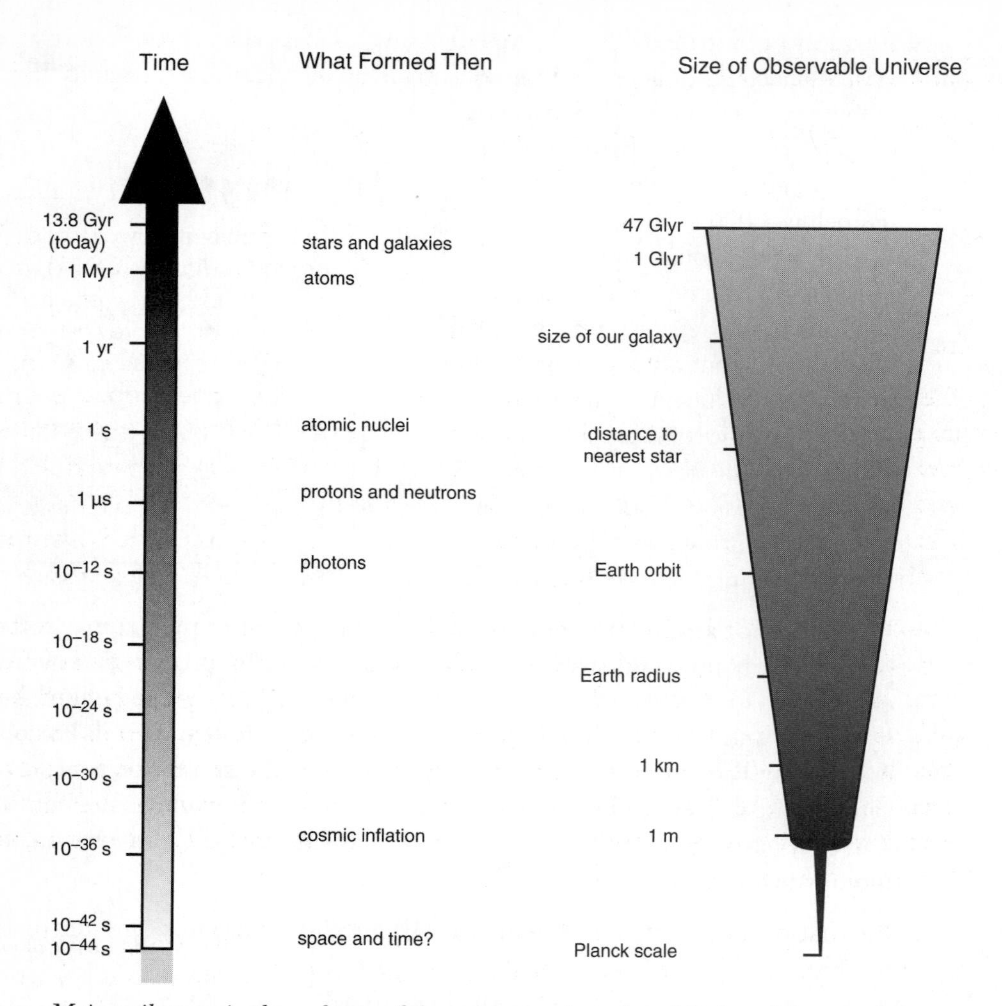

Major milestones in the evolution of the universe, using a logarithmic scale because the pace of events was faster in the ancient universe.

Notice what has happened as we've gone back in time: we've gone down the tree of physics. Only a couple more branches lie beneath us. The biggest particle accelerator, the Large Hadron Collider near Geneva, can recreate conditions as far back as the formation of photons. To carry on to even earlier times, cosmologists have to shift to a different line of argument.

Cosmic Inflation

With all those particles pressed up against one another and plenty of time before the sun would come up, the youthful universe must have been quite a party, if you're the particulate type. But like a lot of human parties, it took a while to get going. Although particles were scrunched together, even a particle needs time to mingle, so at first each one felt utterly alone in the crowd. Only gradually did they begin to interact, creating a small but widening circle of intimates. Meanwhile, the universe expanded in size, making it ever harder for particles to hook up.

The fact that mingling takes time leads to a serious puzzle. If you look at the night sky to your left, it looks a lot like the night sky to your right. The details differ, but the farthest galaxies on the two sides are statistically indistinguishable. If you look above and below you, in front of and behind you, and in thousands of other directions, you also see that the galaxies look about the same. How can that possibly be? The particles in those galaxies never had a chance to mingle. They have never even seen each other.

To switch metaphors, those galaxies are like two ships on opposite sides of our horizon, visible to us but still hidden from each other. Light from the left one has been traveling for 12-odd billion years and is only now reaching Earth. It will need to travel billions more years to reach the galaxy on the right. Ditto for light coming the other way. In short, light, heat, and other influences have not yet had time to pass between them and coordinate their appearance. The puzzling fact that galaxies look the same even though the big bang gives them no chance to harmonize themselves is called the horizon problem.

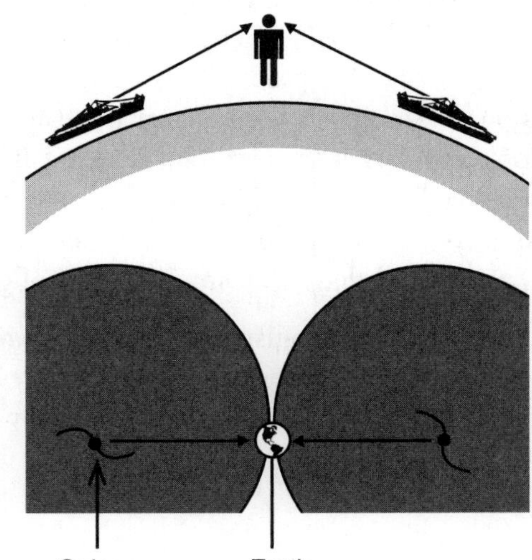

We can see two ships on our horizon even when they can't see each other (top). Similarly, we can see two galaxies on our cosmic horizon even when they can't see each other (bottom).

This riddle indicates that the period of cosmic expansion we're in can't be the full story. There must have been some preparty when the galaxies or their precursors mingled before going their separate ways and meeting up again later. The leading account of this earlier period is a process known as *inflation*. The idea is that the galactic precursors originated in a tiny region of space and had time to rub elbows. Then cosmic expansion went into overdrive and yanked them apart. After a while, expansion mellowed out, and light started to pass between the galaxies again.

In fact, all that's necessary is that expansion mellow out within our region of space. It might well continue in the universe at large, far beyond the range of our vision—an idea known as *eternal inflation*. Somewhere out in that vast realm, another region of space could also call it quits. Because of random variations in the starting conditions, this region could look quite different from ours. This is one of the reasons that cosmologists think our observed universe may be only one among many (see Chapter 15).

def•i•ni•tion

The cosmological variety of **inflation** is an acceleration of the rate of cosmic expansion, pulling neighboring regions apart too fast for them to exchange signals or other influences. The **inflaton** (no "i") is the unknown quantum field and associated particle that brought this about. **Eternal inflation** is the idea that inflation, once started, continues forever in the universe at large. Our observable universe is just a small region where it happened to stop.

Cosmologists attribute inflation to a substance called the *inflaton* that no longer exists in today's universe. Average stuff just isn't up to the task because its gravity attracts other stuff; if anything, it holds the expansion back. To throw cosmic expansion into overdrive takes a freakish type of substance, one that produces a sort of "antigravity" that repels material, giving the expansion an extra push. General relativity allows that. The only trouble is that physicists don't know what the cosmologists are talking about. Nothing in the Standard Model quite matches the description of the inflaton, so physicists need to seek it out in a unified theory—as a byproduct of either the grand unification that marries quarks with electrons or a full-fledged quantum theory of gravity.

The process never ceases to amaze cosmologists. It implies that giant bodies such as galaxies originated in teensy-weensy random fluctuations. Telescopes become microscopes, letting physicists see down to the roots of nature by looking up into the heavens.

Quantum Leap

Whatever the inflaton was, not only did it generate the force that drove galaxy precursors apart, but it also generated the precursors themselves. The inflaton fluctuated in energy as all quantum things do. Under normal circumstances, such fluctuations average out over time, an uptick later being offset by a downtick or vice versa. But the expansionary overdrive was not a normal circumstance. It enlarged the fluctuations just as they formed, preventing an uptick from being compensated by a subsequent downtick or a down by an up. Thus the fluctuations were locked into place like waves on a frozen pond. Later, when the expansion mellowed, the upticks became the seeds for galaxies.

The Ultimate Beginning

Inflation acts as a cosmic reboot. The huge expansion dilutes any pre-existing particles by so much that they are effectively erased. It resets the density of the universe to its cosmological default value and fills it with a fresh batch of particles. When you reboot a computer, a clean start is the goal. But for cosmologists, it poses a problem. How do they probe what came before if it has been almost totally wiped out?

One thing cosmologists do know is that something had to precede inflation, so that the universe could attain the conditions to start its wild ride. Either the universe was born inflating or it had a preinflationary life when it must have been expanding. Therefore, the argument for a moment of complete and total scrunching still applies. Inflation might push this singularity too far back in time to see, but it's hard to get rid of it—in relativity theory, at any rate.

The hope has always been that quantum effects will kick in and prevent the density from becoming truly infinite. The singularity might prove to be the true beginning of time or merely a transition from a pre-existing state, maybe the latest stage in an eternal cycle. Physicists and philosophers have argued it both ways.

Yet current quantum theory can't be the whole answer either, because it works within time. It tells us what will happen given what came before. But that presumes each moment has a moment before it, which doesn't apply to the origin of the universe. Whether or not the universe is finite or infinite in age, there was no "before" in which a process could operate. To ask about "before" is a loaded question, like asking what's north of the North Pole or past the last number on the number line or clockwise of the first point on a circle.

Some scientists argue that if there is no "before," then there is no need to explain the origin of the universe. It just is. Alternatively, maybe what came before simply doesn't matter. Inflation, among other processes, resets the universe to its default values anyway. But these responses run into a practical difficulty. As cosmologists probe back, they find there are some questions for which the starting conditions can't be avoided. One, ironically, is inflation: getting it rolling takes very special conditions. Another is the arrow of time, the distinction between past and future—the fact that we can't remember the future or change the past and that desks get messier but do not spontaneously tidy themselves up. The laws of physics can't explain the arrow because they do not single out a direction in time. The answer must lie in the starting conditions.

This may be another case when relativity and quantum theory cover each other's blind spots. Relativity isn't trapped inside time as quantum theory is. It's a theory *of* time. So it holds out the hope of explaining the dawn of time without presuming time from the outset. Physicists have yet to pull off this feat, though. It's a major motivation for a quantum theory of gravity.

The beginning of time may prove to be the end of time, the point where the limitations of time as a concept become apparent and has to be replaced with something more basic. Just as our human lineage began with something that wasn't even human and the tree of life is rooted in things that weren't even alive, space and time must emerge from things that aren't even space and time.

The Dark Side

Every family has its black sheep, its prodigal son or daughter who runs off to make it big in the city or across the water without so much as a glance backward—except, of course, when they need some money from those who stay at home. Our cosmos, too, has its black sheep. Look in the mirror; we're it.

The matter I've been calling "ordinary" is actually pretty extraordinary by cosmic standards. It's a cauldron of nuclear reactions, shock waves, magnetism, and turbulence. It forms structure on all scales, from protons to people to planets. Most of the universe is not so interesting.

One of the great discoveries of twentieth-century astronomy is that what we see is not all there is. Like a theater audience that watches the herky-jerky gestures of a marionette and infers the presence of a hidden puppeteer, researchers have observed that visible matter moves in unaccountable ways and concluded that unseen matter must be pulling the strings. For instance, stars orbit so fast within galaxies that they'd fly out unless something reined them in. The familiar view of galaxies as big bundles of stars is now passé. Galaxies are really just giant balls of *dark matter* with some stars and gas sprinkled in.

def•i•ni•tion

Dark matter is the unknown material that makes up the bulk of the matter in the universe. It doesn't emit or absorb light but does exert a gravitational force.

Whatever this stuff is, it can't be ordinary stuff. From the measurements of the relative amounts of elements, cosmologists have estimated the total amount of ordinary matter cooked up by the big bang. It falls short by a factor of 10. Starlight and the cosmic microwave background radiation chip in a bit more but not much. Something else must be lurking out there.

Dark matter doesn't behave like the matter we know and love. It doesn't form atoms, doesn't form molecules, doesn't exert pressure like a gas, doesn't respond to pressure, doesn't give off light, and doesn't absorb light. It doesn't do much of anything; it just sits there on the couch all day exerting gravity. If it has any inner life at all, it keeps it well hidden from us. But don't discredit it. By not responding to light, dark matter was able to clump even when ordinary matter, buffeted by the sea of radiation that filled the early universe, couldn't. If not for that, ordinary matter would still be spread out into a thin, lifeless gruel. Dark matter is what gave us our start in life.

Even dark matter, though, has itself become a black sheep in the very cosmos it helped create. It has been shoved aside by *dark energy*, a relative latecomer that has been making up for lost time. About 7 billion years ago, which is recent in cosmic terms, measurements indicate the expansion rate of the universe began to pick up. It's shifting into overdrive, a lot like what happened in the early inflationary period. The cause might be an inflatonlike particle; it might be related to that great bugaboo of quantum theory, the cosmological constant (see Chapter 7); or it might be some new effect altogether that masquerades as energy.

def•i•ni•tion

Dark energy is the unknown type of energy that's causing the expansion of the universe to accelerate.

Either way, it's the first qualitatively new thing to burst onto the cosmic scene since the formation of photons. It now accounts for nearly three-quarters of the energy content of the universe. It's especially strange that dark energy emerged around the same time as star and galaxy formation hit its stride—suspiciously fortuitous timing that cosmologists call the *cosmic coincidence*. Why it lay low for billions of years, waiting for its moment to strike, must be some kind of clue to what it is.

def•i•ni•tion

The **cosmic coincidence** problem is the unexplained fact that dark energy became dominant not long after stars and galaxies began to form.

Dark energy contributed to our existence in the same way that the Mafia gives you a few extra days to pay up. By not barging in until about halfway through cosmic history, dark energy gave galaxies and groupings of galaxies a chance to take shape. The grace period is over now. No larger groupings of galaxies can form in today's universe. Dark energy pulls them apart before they have a chance to assemble.

In sum, the Standard Model of particle physics covers only about five percent of the universe: the stars, intergalactic gas, radiation, and sundry particles. The remaining 95 percent of the universe will take a unified theory to identify. The names cosmologists give it, dark matter and dark energy, are really a fancy version of "hey, you!" and "that guy over there." Ten years from now, if experimenters are able to manufacture dark matter in particle accelerators and theorists are able to explain it, it will have a real name, and we will finally be reunited with our long-lost relations.

The Least You Need to Know

- Going back in time means diving deeper into the workings of nature.
- Cosmic expansion could enlarge strings to astronomical size.
- The cosmic microwave background radiation provides a crucial set of observations.
- Only quantum gravity can make sense of the dawn of time; it might also be needed to tie up other loose ends, such as inflation, dark matter, and dark energy.

Chapter 10

Time Machines

In This Chapter

- Wormholes and building a time machine
- The role of negative energy
- The grandfather paradox and others
- Time travel meets quantum gravity

Time machines are a different kind of case study from black holes or the mysteries of cosmology. Those phenomena exist, and a quantum theory of gravity seeks to explain why. Time machines *don't* seem to exist, and a quantum theory of gravity seeks to explain why not. Only such a theory will be able to decide between general relativity, which says we can travel into the past, and common sense, which suggests that time travel pose too many riddles to be possible.

Blueprint for a Time Machine

Time travel into the future is uncontroversial: it's called life. All we can do is grip the wheel tightly. But wouldn't it be nice to have some control? Einstein's theories of relativity say we do.

His first theory, special relativity, says that the throttle of a moving vehicle, be it a plane or a rocket, is also the throttle of time. The faster we move, the slower time passes for us compared to other people (see Chapter 3). On a New York–London airline flight, we age about 30 nanoseconds less than our brother who stays at home. It's not much, but it's enough to throw atomic clocks out of sync unless scientists compensate for it. At higher speeds, approaching that of light, we could age years less than the person who stayed home. Astronauts on a short but swift space voyage could return to Earth only to discover that thousands of years had passed and the human race had gone ape. No one has ever done such a trip for real, but astronomers have observed fast-moving subatomic particles that crossed the entire galaxy in what was, for them, the blink of an eye.

So we all move into the future at our own pace. Because time passes differently for different people, the distinction between the past, present, and future must be subjective. What counts as "now" for me could lie in your past or someone else's future. For all of us to be right, all moments in time must exist equally. Most (though not all) physicists think time is laid out in its entirety like a landscape, and what we perceive as past, present, or future depends only on where our gaze happens to fall. The past is not past. It still exists and always will—and maybe we could visit it.

Einstein's second, more advanced theory, general relativity, goes further and says that the past is accessible to us. If spacetime is malleable, we can bend it like a glassblower to create a passage between far-flung locations in space or moments in time. The theory dictates how each patch of spacetime curves but is lenient about how the patches are sewn together into an overall shape. There's no shortage of shapes that allow for time travel.

In the Loop

General relativity is completely infested with time machines.

—Matt Visser, Victoria University of Wellington

Wormholes

The best-studied hypothetical time machine is a wormhole, a shortcut between two points in spacetime. Just as a tunnel passing under a hill can be shorter than the surface street, a wormhole may be shorter than the normal route through ordinary space. If a wormhole connected New York and London, we could travel between the cities with a single step. Someone who didn't know about it would have to sit seven hours on a plane (not counting the seven hours passing through security checks). This wormhole is not a tunnel through the rock between the two cities, which would shave only 100 or so miles off the journey anyway. The wormhole is a brand-new connection in spacetime. Via this connection, the two cities can be literally just a few feet from each other.

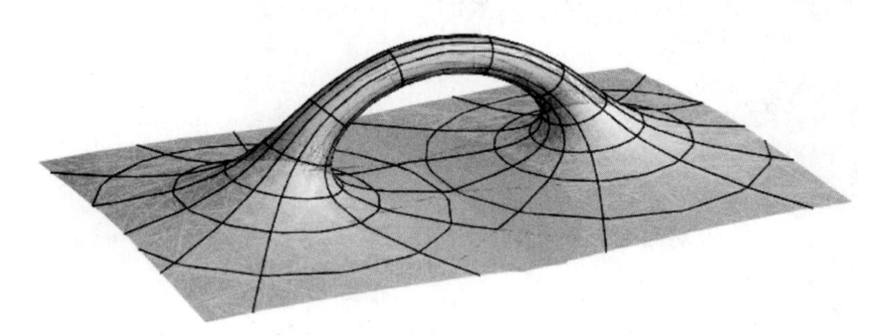

A wormhole is a tunnel across spacetime, creating an alternate path between two points.

If we could venture outside spacetime and look at the wormhole—an impossible view, but one that helps to visualize what's going on—it would look like an archway. The gap between the archway and the main region of space is the defining characteristic of the wormhole. It ensures that the two paths from New York to London remain completely distinct, unable to merge into a single path even if the detailed shape of the wormhole or of the main region changes. Conversely, the wormhole can slide around without greatly affecting the shape of the rest of space.

Those of us stuck inside spacetime don't see any arch. We just see wormhole entrances, or "mouths," which resemble crystal balls. Peering into the New York–London wormhole from the American side, we could see Big Ben. Someone on the other side would see our smiling faces looking at them. It's like a Skype video-chat window except that we could actually reach through the screen and touch the other side.

Although I've described this wormhole as connecting two cities, we could use it to connect any two points, including different points in time. To use a wormhole as a time machine, the first and most difficult step is to find one. Relativity theory forbids us from making one ourselves because that would require tearing and reconnecting the fabric of spacetime (a real no-no), but one might have been created during the formation of the universe and been locked in place ever since. The curved spacetime around each mouth would give it away: light rays passing nearby would bend in a distinctive way. Astronomers have looked for wormholes, so far without success, but let's imagine that they locate one someday.

Through a wormhole mouth, we'd see a distant location—in this case, an office.

(M. C. Escher's "The Sphere." Copyright 2007 The M. C. Escher Company–Holland. All rights reserved. www.mcescher.com)

The next step is to move it into place. A spaceship could tow one of the mouths to another location. Recall that time passes more slowly for a moving object. The wormhole mouth is no exception; as it moves, time slows down for it. Consequently, the two mouths become separated not only in space but also in time. Then we can bring the two mouths back together at a slower speed. The time difference between them will remain locked in.

Towing a wormhole is a technologically advanced, but otherwise fairly benign, operation. The wormhole mouth would probably be quite narrow, and as long as we keep pets and small children out of its way, we can drag it without leaving a trail of destruction. Nor does the towing operation change the internal length of the wormhole, which depends only on the arrangement of matter inside it. All towing does is reposition the mouth within ordinary space.

At last, we can go through the wormhole. Depending on which mouth we enter, we travel into our own past or future. This kind of time machine doesn't have a dial allowing us to specify the time we want to go to. It's just a fixed difference: if we enter one mouth, we come out the other a certain number of years later or earlier.

All Tangled Up

If time machines were possible, shouldn't we already have one, sent to us as a gift from some kind traveler from the future? Actually, all plausible designs for a time machine prevent that from happening. They only allow travel to times after they were first built. If we set up our wormhole on January 1, 2010, the farthest back we could ever go is January 1, 2010. We couldn't travel back and give ourselves the machine before we'd made it. Sorry, but the inventor of the world's first time machine is going to have to build it from scratch.

None of this is about to happen anytime soon. A physicist's job is to sketch out what is possible, as a way of probing the laws of nature. Engineers have to figure out whether it's practical.

Negative Energy

One potential roadblock is that a wormhole is inherently unstable. It wants to pinch off into a point of infinite or near-infinite density—in short, to morph from a wormhole into a black hole. To prop it up, we have to inject some energy that provides a gravitational repulsion, counterbalancing the tendency to collapse. That's easier said than done. Ordinary energy exerts gravitational attraction, so to exert gravitational repulsion, we need some very unordinary energy. It must be negative—an amount less than zero. That means we empty out a box, leaving a vacuum, and then continue to empty it, leaving literally less than nothing.

Hey, what did you expect from a time machine? That you could build it out of papier-mâché? A weird contraption takes weird stuff. Note that negative energy is not the same as antimatter. When matter hits antimatter, they annihilate in a bang, leaving behind a blaze of radiation. If a particle with negative energy hits one with positive energy, they annihilate with a whimper, leaving behind zip, zilch, nada.

All proposed time machines, including those that don't involve wormholes, either demand negative energy or have some other, equally daunting requirement. At this point, relativity reaches the end of its rope. Negative energy doesn't exist in classical physics.

But it does exist in quantum physics, providing a first hint that understanding time machines requires the merger of relativity and quantum theory. Quantum theory says that the vacuum is a dynamic place, filled with fluctuating fields. If you can damp down these fluctuations, the vacuum will have less energy than usual—that is, less than zero energy. That is doable. One approach is to fight fire with fire. You can cancel out the fluctuations by creating offsetting fluctuations, just as noise-canceling headphones create offsetting waves that meet and cancel the outside noise. Another approach is a sort of quantum guillotine. A pair of parallel metal plates, which chop off the fluctuations in the region between them, can reduce the energy below zero.

Quantum Leap

Not only do quantum processes let us stabilize a wormhole, they ease the task of locating one to begin with. Although relativity forbids us from tearing and reconnecting space, quantum fluctuations might evade this restriction and create wormholes of subatomic size. You could conceivably pluck one out of the vacuum and enlarge it to human proportions using a quantum field similar to the one that caused the universe to inflate very early in its history.

However, there are limits. To create a shimmering blob of negative energy, you must extract energy from the vacuum, and you're eventually going to have to give it back again. Negative energy is basically an energy loan. Just as a debt is negative money that has to be repaid, negative energy also has to be repaid. What's more, you have to repay it with interest: the amount of energy you restore to the vacuum must overcompensate for the amount you extracted, leaving behind a net positive amount. In other words, you can't get something for nothing, even if the something is less than nothing.

This restriction doesn't seem to be a show-stopper. Physicists used to think that a useful wormhole would require producing an unrealistically large amount of negative energy—equivalent to the mass of a large planet, say—in an unrealistically tiny volume of space. But some now think that small amounts might do the trick.

What's Wrong with Time Machines?

Time-travel paradoxes are to science fiction what complex, drawn-out courtships are to Victorian novels. On close examination, none of these paradoxes is really all that paradoxical. The most popular one is that we go back in time and alter the past in such a way that it prevents us from going back in time. It usually goes by the name of the "grandfather paradox," because of the morbid thought that you could go back in time, shoot your grandfather as a young boy, and prevent yourself from ever being born—in which case you couldn't go back in time and kill him, in which case you *could* go back, in which case you couldn't. You get yourself caught in an endless loop of contradiction. The paradox works with other familial relationships, too: your grandmother, your mother, your earlier self, your whole civilization.

It isn't really a paradox, though. It doesn't prevent you from traveling into the past; it just means that if you did try to kill your grandfather, something or other would stop you. It has to. The real source of the paradox isn't the time machine, but the notion that we are complete masters of our destiny. We're not. We can't do anything that violates the laws of physics, and preventing yourself from existing falls into that category. Another way to think about it is that there is only one past. How could you change it? If you go back into the past, you're part of it. In science-fiction stories, what typically happens is that characters who change the past end up creating the very situation they wanted to avoid.

Quantum Leap

Quantum theory, being based on probability, adds a provocative twist to the resolution of the grandfather paradox. The key is that events can occur with a certain probability. Let's look at it in chronological order. You, the time traveler, walk out of a wormhole with a 50 percent chance. If you do, Grandpa dies. If not, he lives. The outcome: a 50–50 chance of your being born and later entering the time machine. So probability makes the whole story consistent.

Another apparent paradox of time travel is the "bootstrap paradox," whereby the time machine conjures something into existence. For example, a time machine could act as an all-powerful computer, able to provide the answer to any question that had an answer. The computer works simply by waiting for the answer to pop out of the time machine. Once it has the answer, it sends it back through the time machine to its earlier self. It's the ultimate cheat: the computer manufactures information out of nothingness. It's shady, but logical.

Although the paradoxes aren't strictly paradoxes, something about them still bothers most physicists. What exactly prevents you from killing your grandfather if you try? There are many scenarios—too many. You might fumble the gun; the gun could misfire; another person could materialize out of the time machine and knock the gun out of your hands; and so on. Which will it be? The laws of physics can't decide, which is a serious failing. Given the initial setup, the laws should always be able to predict what happens. With a time machine, they can't. Almost anything could happen.

As for the bootstrap paradox, it allows an effect to precede or even become its own cause—scrambling a sequence that relativity theory was supposed to preserve. If the distinction between cause and effect breaks down, children would give birth to their own parents; criminals would be punished before committing their crime; and students would get their grades before taking their tests. The world becomes a total muddle.

The Role of Quantum Gravity

Because time machines do such violence to the principle of determinism and the distinction between cause and effect, the fact that relativity theory permits them strikes physicists as just wrong. They conclude that the theory must be missing something, not that time machines are really possible. But they can't quite put their finger on what that something might be.

Some, such as Stephen Hawking of Cambridge University, have proposed that a time machine sows the seeds of its own destruction. Photons of light are always passing through the machine. In so doing, they can pick up energy. For example, if the wormhole mouths are moving toward each other, a photon gets a boost as it exits one mouth and then re-enters the other. As the photon loops through the machine, it picks up ever more energy—without limit. This feedback loop is like the squeal of a microphone held close to a speaker: sound goes in the microphone, out the speaker, back into the microphone, and soon everyone is covering his ears. If the photon starts off as visible light, it turns into ultraviolet light, then x-rays, then gamma rays, and beyond. Not only does it become a serious radiation hazard for anyone near the machine, but also its energy grows so enormous that it warps the spacetime around it. Presumably the wormhole would collapse.

Other physicists think the feedback loop could be brought under control. For instance, the curved spacetime around the mouth of the wormhole tends to defocus light—that is, cause it to spread out. This reduces the energy intensity and can squelch the feedback, just as turning down the gain on the microphone stops that ear-splitting squeal.

The difficulty of identifying a restriction without an escape clause suggests that the answer must lie in a full quantum theory of gravity. Quantum gravity kicks in even before the time machine becomes fully operational. Let's look a bit more closely at how photons circulate through the wormhole. While the spaceship is still towing the wormhole mouth into place, a photon can enter one mouth, come out the other, and then travel through ordinary space back to its point of departure. It arrives after it had left. Once the wormhole becomes a time machine, the same photon arrives before it had left. After all, that's what it means to be a time machine.

But at the instant when the wormhole becomes a time machine, the photon arrives exactly when it had left. Suppose the time difference between the mouths is one year and the two mouths are located one light-year apart. A photon enters one mouth, comes out the other a year earlier, and travels back through ordinary space, a journey that takes it one year. It returns home at its precise moment of departure.

A split second earlier, the photon arrives within a Planck time of its departure. Whenever we see "Planck," we know that quantum gravitational effects come into play. On Planck scales, distance fluctuates unavoidably, and the photon feedback loop amplifies those fluctuations, making them observable. At that stage, all bets are off. It will take string theory or another such theory to determine whether the wormhole (let alone Grandpa) survives or not.

Until we actually have time machines, their main benefit is conceptual. Like black holes, they are an important test for a quantum theory of gravity. Most physicists hazard a guess that if a proposed theory allows for time machines, it's probably wrong. Disappointed? I am, too. Most of us would give anything to have a time machine and be able to have a do-over for the mistakes we've made. But maybe we should be careful what we wish for. A time machine would be nothing but frustration. We'd see all the errors of the past and be utterly unable to rectify them. Maybe we're better off leaving our grandfathers in peace.

The Least You Need to Know

- General relativity allows time travel.
- Time travel involves paradoxes, but none is a show-stopper.
- Physicists worry that a time machine jumbles cause and effect.
- Settling whether a time machine is possible will require quantum gravity.

Part 4

Gravity Meets the Quantum

Unifying relativity and quantum theory is a bit like the case of the irresistible force meeting the immovable rock. Both theories are based on compelling principles, but they can't both be completely right because contradictions arise when we try to unite them. It's hard to know which needs to give way. Decades of effort have converged on two main approaches, one growing out of particle physics (string theory) and the other out of relativity (loop quantum gravity), as well as other contenders that are still very much in the running. String theory is the choice of the majority of physicists, but loop gravity and the other alternatives have important insights as well.

Chapter 11

The Paradox of the Graviton

In This Chapter

- Putting the "quantum" in quantum gravity
- Introducing the graviton particle
- Going for a spin
- Why points pose problems
- Understanding the gravity of the situation

How hard could it be, really, to make a quantum theory of gravity? After all, quantum theory manages to explain the other forces of nature just fine. Why should gravity be any different? Well, try to force-fit gravity into the quantum framework and the framework will buckle.

The Primacy of Quantum Theory

The very term "quantum gravity" implies that unifying quantum theory with relativity theory won't be a merger so much as a takeover. Quantum theory gets to keep the corporate logo, while relativity becomes just one of the subsidiaries. Where the two conflict, relativity will be the one to give way.

The reasoning is simple and persuasive. Quantum theory is the most successful theoretical framework in the history of the physical sciences. It can do everything prequantum theories can and so much more. Those prequantum theories work well for the situations humans typically encounter, but once we venture into new settings, they run out of steam. They just can't handle the range of phenomena that quantum theory can—not only exotic ones, such as superconductors and quantum computers, but also humdrum ones, such as the fact atoms don't implode. Quantum theory has subsumed all the theories that preceded it, with the lone exception of general relativity.

Einstein's theory has to join the party sometime. In it, gravity arises from the presence of matter and energy. Matter and energy are inherently quantum, so it makes sense for gravity to be, too. Quantum fluctuations in matter and energy should naturally give rise to fluctuations in the gravitational field, which will take a quantum theory of gravity to describe coherently.

So far, relativity has held out because the tension becomes unbearable only in extreme settings, such as black holes. But eventually it'll have to find a way to fit in. Though not strictly "wrong," it may prove to be an approximation to a deeper theory, or it might just need to be reformulated.

Meet the Graviton

If quantum theory supersedes relativity in the same way it superseded other prequantum theories, then gravity, like the other forces of nature, must be transmitted by a dedicated type of particle. Physicists in the 1930s dubbed it the *graviton*. It bears a certain resemblance to a little-known particle you have created in your living room: the carpeton.

When you shove on one side of a carpet, you create a ripple in the fabric. As you try to smooth it out, it slithers across the room, shifting the position of the coffee table, plastic toy lizards, and anything else lying around. If a remote-control car is driving across the rug, the ripple pushes it off course. A carpeton is the gentlest possible ripple you could make. Add enough of them up, and you can get a wave capable of toppling the furniture.

def•i•ni•tion

The **graviton** is the hypothetical particle that transmits the force of gravity. It is the smallest unit of a gravitational wave.

A **gravitational wave** is a ripple in the gravitational field and therefore in the fabric of space and time.

Gravity is nothing more or less than a distortion of the spacetime fabric. Altering the gravitational field exerted by a body creates a *gravitational wave*, which takes on a life of its own and spreads out through spacetime, reshaping it. Distant bodies, gliding along the contours of spacetime, shift onto new paths. In quantum theory, the wave consists of gravitons. Celestial motions typically involve the production of billions upon billions of gravitons, each causing such a small effect on its own that you can ignore the individual particles and focus on their cumulative effects, which is what general relativity describes.

Physicists model the graviton on its electromagnetic counterpart, the photon. Both are generated when something disturbs their respective fields, gravitational or electromagnetic. For instance, a radio transmitter creates photons by pushing or pulling on electrically charged particles, typically electrons. A gravitational wave transmitter creates gravitons by pushing or pulling on particles of any sort. For gravitons, the role of electric charge is played by mass and energy. The more massive a particle or other object, the more gravitons you create by wiggling it.

Quantum Leap

Want to create some gravitons? Get working! You'd have to wave your arms once a second for a million years to generate enough energy for a single graviton. The orbit of Earth around the sun produces about as much energy in gravitons as a single light bulb produces in photons. Serious graviton production takes a tightly orbiting pair of stars.

According to prequantum theories, if an object is sitting still and not disturbing its field, it doesn't produce waves of any sort. After all, if nothing changes, nothing needs to be transmitted. A comb with a static electric charge doesn't need to keep beaming electromagnetic waves at your hair to keep it standing on end. Once the waves sculpt the electric field around your hair, their job is done; the field is what keeps it levitated.

Quantum theory sees things a little differently. It describes forces not as a field that gets locked into place, but as a continual exchange of "virtual" photons even in the absence of a disturbance. Most physicists think that gravitons work the same way. According to general relativity, the sun doesn't need to keep beaming gravitational waves at Earth to keep it in orbit. The spacetime fabric, once bent into shape, guides our planet. In the quantum version, gravity is conveyed by a continual exchange of virtual gravitons.

Despite their many similarities, gravitons and photons have conspicuous differences that get to the heart of what makes quantum gravity so much harder than quantum electromagnetism. For starters, whereas sensitive detectors can spot individual photons, they don't have much hope of ever picking up individual gravitons. Even trying to catch them en masse pushes modern technology to its limits (see Chapter 21). So gravitons will probably always be hypothetical. Even so, they will play an important conceptual role by helping to explain the harmony of nature.

Putting a Spin on It

A deeper difference between gravitons and photons arises because electric charges come in two types: positive and negative. Photons affect them differently, ensuring that like charges repel and unlike charges attract. Gravity, on the other hand, affects all bodies in the same way. Matter, antimatter, positive energy, negative energy, Democrats, Republicans—makes no difference. Drop them all from the Tower of Pisa, and they'll all hit the ground at the same time. They're falling freely through the same patch of space, so nothing differentiates them.

Consequently, when gravitons pass through a material, they must be completely even-handed. They can stretch the material horizontally as long as they squeeze it vertically to compensate. A moment later, the graviton squeezes where it had stretched and stretches where it had squeezed. Depending on the thickness of the material and on the wavelength of the graviton, it could go through many cycles of squeezing and stretching, until finally exiting and leaving the material as it was.

This gravitational kneading remains the same even if we rotate the wave by 90 degrees. Over a full cycle of the wave, there's no difference between horizontal and vertical: the graviton alternately squeezes and stretches in both directions. In contrast, we have to rotate an electromagnetic wave by 180 degrees to keep its effect the same. If we rotate by just 90 degrees, the wave will cause electric charges that had been oscillating left and right to start wobbling up and down instead. In this sense, gravitons have the symmetry of a plus (+) sign (a quarter turn keeps it the same) whereas photons have the symmetry of a negative (-) sign (only a half turn does). So they are twice as symmetrical as photons. In the jargon, the photon is spin-1, whereas the graviton is spin-2.

This might sound like a minor change, but it makes a big difference. For one thing, we can flip the argument around: if we come across a spin-2 particle, we have good reason to think it produces a force that treats all bodies equally—a force none other than gravity. The extra symmetry of the graviton requires the complex mathematics of general relativity to describe. This is how it dawned on physicists that string theory might be a quantum theory of gravity. They developed the theory for other purposes, discovered that it predicts a spin-2 particle, and realized it might be the long-sought quantum theory of gravity (see Chapter 12).

A gravitational wave passing through this page would first stretch it from left to right while squeezing it from top to bottom, and then switch.

Caught in an Infinite Loop

Albert Einstein spent the last third of his life working on a unified theory of physics, based on the idea that the similarities between electromagnetism and gravitation mattered more than their differences. However, this turned out not to be the case. Electromagnetism is actually more closely related to nuclear forces, which at first seem so unlike it.

Electromagnetism and nuclear forces have the special property that they are scale-invariant (see Chapter 6). This means that no matter how closely we look at a particle or how extreme the conditions become, electromagnetism acts in essentially the same way. Its strength does vary somewhat, but this doesn't reflect a fundamental change in behavior; a slight adjustment of the electric charge and the masses of charged particles fully captures this variation. If someone showed you a picture of a magnetic field, you'd be hard-pressed to tell whether it was produced by a star, a planet, a refrigerator magnet, or a magnetite nanoparticle.

Gravity, on the other hand, is scale-dependent. Unlike electric charge, gravitational "charge"—namely, mass and energy—varies hugely with scale. As we turn up the microscope magnification or smash particles together at ever greater speeds, gravity not only gets vastly stronger but also behaves in qualitatively new ways.

The effects of scale become evident when we try to predict how photons or gravitons pass between two bodies. In quantum theory, everything not forbidden is compulsory, and there's nothing to stop the particles from spontaneously splitting and reuniting. So they do, with some probability. The humblest exchange of a photon or graviton triggers a mighty cascade of them. For photons, though, the splitting follows a very rigid pattern. Whatever the energy of the photon, it splits and reunites in the same way; the interaction is merely scaled-up or scaled-down in size.

The situation is like measuring the length of a coastline or a tree. The closer you look, the more rugged it gets, but each level is just a miniaturized version of the previous one. In principle the endless progression of zigs and zags adds up to an infinite length, but knowing the trend, we have no trouble measuring the length to any desired precision. When we get down to the atomic level, the calculation presumably fails, but until then it works just fine. Similarly, we can calculate the strength of a photon whatever its energy.

Dealing with the exchange of gravitons is far harder. Because of gravity's scale-dependence, the pattern of splitting and reuniting is like a coastline or tree that gets increasingly rococo as you zoom in. We must assume that some outrageously baroque things befall the gravitons on their journey, and the more extreme the conditions, the more baroque things we need to postulate. The gravity we observe under ordinary circumstances is no guide to what happens under extraordinary circumstances.

In short, the very theory that introduced the graviton is unable to describe it fully. It predicts that even the most routine exchange of a graviton becomes infinitely complex, so gravity becomes infinitely strong.

A scale-invariant tree has the same shape at all scales (left), whereas a scale-dependent one gets progressively more complicated (right).

(Courtesy of Richard Taylor)

One response is to revamp the theory—that's what string theory does. It supposes that general relativity must be an incomplete theory of gravity. The other is to give up trying to build up a theory particle by particle—that is loop quantum gravity's response. It supposes that gravity can't be incorporated into quantum theory using the same techniques that work so well for the other forces.

There's Too Much Room at the Bottom

The fact that gravity is scale-dependent is a case of good news/bad news. The Planck scale appears in neither quantum theory nor general relativity on its own. Each of these theories treats spacetime as a smooth continuum. In principle, there's no limit to how closely we can zoom in. Spacetime is like a computer screen with infinite resolution; its pixels are true mathematical points, having zero size.

Any time the number zero appears, the number infinity can't be far behind. If particles interact at points of zero size, don't their properties have to change with infinite speed? How could a point have internal properties such as electric charge when the concept of "internal" has no meaning for it? Any region of spacetime consists of an infinite number of points, requiring an infinite amount of information to account for all the possible goings-on, and isn't that just plain silly?

In the Loop

It always bothers me that, according to the laws as we understand them today, it takes a computing machine an infinite number of logical operations to figure out what goes on in no matter how tiny a region of space, and no matter how tiny a region of time. How can all that be going on in that tiny space? Why should it take an infinite amount of logic to figure out what one tiny piece of spacetime is going to do?

—Richard Feynman, Caltech

Interestingly, although both theories raise these questions, their union doesn't. Although the theories let us zoom in as closely as we want, they do so in opposite ways. In quantum theory, we dial up the energy to zoom in, whereas in general relativity, we turn it down. These contrary tendencies intersect at the Planck scale, and the zoom mechanism jams. The Planck scale appears to be the shortest distance that has any meaning, setting a limit to how finely subdivided spacetime can be. Spacetime may not, in fact, be a smooth continuum.

So the good news is that quantum theory and general relativity yet again plug each other's gaps. Their merger could also help tame the graviton as well as the wild horses of the Standard Model. Just as atoms keep the length of a coastline from going infinite, because the zigs and zags of the coastline can't be any smaller than the width of an atom, the Planck scale stops particles from splitting endlessly.

Now here's the bad news. There's a good reason that current theories treat spacetime as a continuum: it keeps special relativity happy. When events occur at specific points in space and time, everybody agrees on the sequence of cause and effect. But if events are spread out, even over a distance as short as the Planck scale, people might disagree, potentially leading to the same paradoxes as time travel does. For instance, we could arrange things so that an effect becomes its own cause. In trying to marry quantum concepts with general relativity, a matchmaker will have to keep Einstein's other theory from getting jealous and walking out.

When the Ground Comes Alive

The above problems are aspects of a deeper issue. Calling gravity a force between two bodies is like calling an earthquake a strange rattling noise in the china cabinet. The shaking of a tremor is only one consequence of what's reshaping the entire landscape, which can have much more dramatic effects, like making Los Angeles a suburb of San

Francisco. Similarly, the force of gravity is only one consequence of what's reshaping the landscape of spacetime, which has much broader effects.

The theory of gravitons doesn't capture those effects. It can't. Quantum theory is based on a conception of spacetime that goes back to Newton: a fixed scaffolding that pinpoints where and when things are. The world according to quantum theory evolves moment to moment, and the pace of events is the same everywhere. Different people may have different time conventions, but each assumes that the passage of a second in one place means the same as the passage of a second somewhere else. It's as if the universe had a master clock.

This master clock keeps quantum theory consistent. At any given instant, a particle can be located in one of many possible locations, each with a certain probability. The master clock ensures that we can add up all those probabilities so that they always come to 100 percent. We may not know exactly where the particle is, but it has to be somewhere.

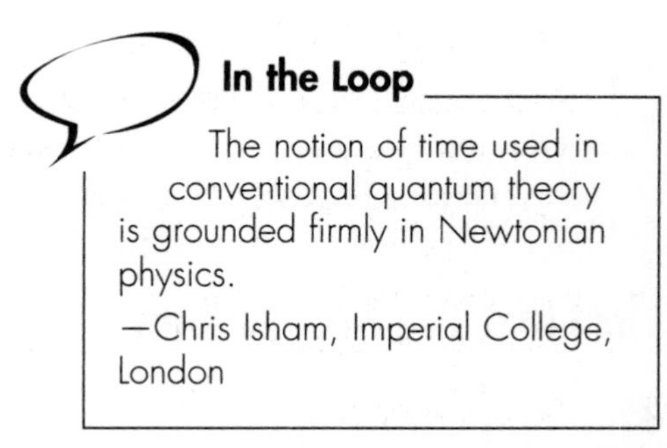

In the Loop

The notion of time used in conventional quantum theory is grounded firmly in Newtonian physics.

—Chris Isham, Imperial College, London

General relativity tosses that clock out the window. The pace of events varies from place to place. A second in one place doesn't mean the same as a second in another. In some situations, it's impossible to tell what is space and what is time, let alone set up a master clock. Nor is there a fixed number of locations where a particle could lurk. For instance, the expanding universe creates new space.

A graviton does this, too, on a lesser but still significant scale. All particles travel through space and time, but the graviton also modifies the space and time it's traveling in. It's a ripple *of* rather than *in* spacetime. So the graviton alters the framework that its own existence depends on. Going back to carpetons, if you jerk your carpet strongly enough, you don't just create a little bump. You fold the carpet on itself, flip it over, or wrap it around your grandmother. Where, in that case, is the carpeton? There's no longer a single bump—even a big one—you could point to. The whole concept of a carpeton breaks down. Similarly, if a graviton is strong enough, how can we even talk of a particle propagating through space? Where is it? Where is it going?

Physicists say that the graviton theory is background-dependent, meaning that there's a fixed spacetime (the "background") that we can always use to situate the graviton. Ultimately, a quantum theory of gravity needs to be background-independent, describing spacetime as fully malleable. String theory and loop gravity take very different approaches to achieving this goal. Loop gravity aims to be background-independent from the outset, whereas string theory starts off as background-dependent and then relates what happens on different backgrounds.

Gravitons raise some of the same issues that black holes (see Chapter 8), the big bang (see Chapter 9), and time travel (see Chapter 10) do. When physicists try to fit the relativity peg into the quantum hole, one victim is the principle of determinism, the concept that everything that happens, happens for a reason. Given the state of the world at one time, the theories claim to be able to predict the state of the world any time later. But how can we predict the state of the world later if we can't even consistently identify "later"?

So the central insight of general relativity, a dynamical spacetime, pulls the rug out from under quantum theory. Either we shouldn't take general relativity completely seriously, or the merger of the two theories needs to be a merger of equals after all.

The Least You Need to Know

- Physicists think gravity is transmitted by hypothetical particles called gravitons.
- Gravitons resemble other force-transmitting particles such as photons, but they have crucial differences.
- The behavior of gravitons keeps changing as we probe more to finer scales until we need a whole new theory.
- This problem reflects the fact that gravitons affect the very space through which they travel.

Chapter 12

The Music of Strings

In This Chapter

- A brief history of strings
- Defining strings
- Strings generate gravity
- The function of branes

At last, the time has come to get tangled up in string theory, the leading proposal for a quantum theory of gravity and indeed for a unified theory of all of physics. It interprets the multitudes of subatomic particles as a single basic unit, the sub-sub-subatomic string, vibrating in different ways.

To Do Is to Be

To the endless consternation of grammar purists, English speakers have a habit of converting nouns into verbs and vice versa: to party all night, to google someone you met there, to cook a romantic dinner for her, to (unfortunately) experience a disconnect. Linguists have a word for this; it's anthimeria. Since when is the distinction between nouns (what things are) and verbs (what things do) all that great to begin with?

Modern physics elevates this habit to high principle. To our eyes, things are primary. They have color, heft, texture, solidity, and lots of gooey stuff on the inside. But as we dig in, their seemingly fixed properties turn out to arise from the actions and interactions of their components. For example, most of us blame the mass of our bodies on inactivity, but actually the mass comes from hyperactivity: the way that quark and gluon particles buzz around inside protons and neutrons, like trillions upon trillions of miniature beehives. The gluons themselves weigh nothing, and quarks hardly anything. It's what they do, rather than what they are, that accounts for the bulk of atoms' mass.

String theory runs with this idea. In fact, it makes nouns almost superfluous. Explaining all the particles of the Standard Model requires just one noun: a string, a teeny little cord. It has only one intrinsic attribute, its tautness. Everything else about strings comes from how they act. Strings are as strings do.

From this deceptively simple concept unfolds a theory of amazing richness. Despite its name, string theory is more of an approach, a set of ideas and tools, than a full-fledged theory. Physicists have found multiple consistent ways to describe strings, each an approximation that applies under certain conditions. What the exact theory is or whether it even exists, no one yet knows. String theorists call this ultimate formulation "M-theory," but the "M" doesn't stand for anything in particular—deliberately. It is just a stand-in for ignorance.

Like the Standard Model, string theory combines quantum principles with Einstein's special theory of relativity. All it does differently is apply these ideas to strings rather than to ordinary particles. Notice that I didn't include the general theory of relativity in this list. That's because string theory doesn't assume general relativity from the outset; it derives it. This singular achievement has kept string theorists going through thick and thin.

A Tangled Tale

Even by Hollywood standards, string theory has an improbable story. Physicists fell in love with it, then broke up with it, made up with it, lost the passion for it, and then found it again. The first string theory was proposed in 1926, right in the swirl of the quantum revolution, and forgotten. (Few string theorists today know about it.) The modern theory had to be reinvented—or, more precisely, stumbled upon.

In 1968 Italian physicist Gabriele Veneziano at CERN (the European particle physics lab) took an equation derived for completely unrelated reasons and noticed that it could explain particle reactions involving the strong nuclear force. Several other physicists—Leonard Susskind, then at Stanford University, Yoichiro Nambu of the University of Chicago, and Holger Nielsen of the University of Copenhagen—realized the equation made sense if they thought of particles as being connected by little strings. This concept was controversial from the start, and within a few years, the Standard Model had won physicists' hearts, leaving the stringy explanation of nuclear particles in Nowheresville.

In the Loop

So I wrote up the manuscript … "Well," they said, "This paper is not terribly important, and it doesn't predict any new experimental results, and I don't think it's publishable in the *Physical Review.*" Boom! I felt like I had gotten hit over the head with a trashcan … I not only got drunk but I passed out and one of my physicist friends had to pick me up off the floor and take me to bed. That was tough. It was not a nice experience.

—Leonard Susskind, Stanford University

A few still saw something in the theory, though. In a gutsy move worthy of *Extreme Makeover*, physicists John Schwarz of Caltech and Joël Scherk of the École Normale Supérieure in Paris rebranded it as a theory not only of the strong nuclear force but of *all* forces—including the most taciturn of them all, gravity. For the next 10 years, this newly expansive theory lived a lonely life. Some of those who stuck with it were denied tenure by their universities and had to scrounge for jobs. Finally, however, they demonstrated that the theory cohered, and a series of conceptual breakthroughs in 1984 fired the shots in what became known as the First String Revolution. By this point, Edward Witten, a renowned particle theorist then at Princeton University, had gotten involved. In a matter of weeks the theory flipped from outsider to insider status. Physicists who had once derided it dropped everything to work on it. It helped that other approaches to a unified theory were going nowhere fast.

Within a few years, the revolution petered out, as revolutions often do. The theory splintered into apparently incompatible versions, and string theorists were strung out. Many physicists switched their focus to getting ready for the Superconducting Super Collider, a giant accelerator being built near Dallas. Then two developments, one negative, one positive, restored string theory to prominence. First, the U.S. Congress pulled the plug on the new accelerator and slashed particle-physics funding. Pure theory began to look like an attractive career choice again.

Second, on March 14, 1995, Witten gave a lecture at the University of Southern California that those in attendance still remember as though it were yesterday. He argued persuasively that string theory's multiple versions were not, in fact, incompatible. This insight re-energized the field and launched the Second String Revolution. It, too, burned brightly for a while.

Some physicists think a burst of developments five years ago, which included a new framework for understanding the shape of spacetime, qualifies as a third revolution, but others wouldn't go that far, and if we have to argue over whether it's a revolution or not, it probably is not. Today, theorists' attention centers on the Large Hadron Collider and other instruments, which could provide some experimental guidance on whether they're heading in the right direction at all.

What Are These Strings, Anyway?

According to string theory, if we magnified an elementary particle enough, we'd see a string—either a closed loop like a rubber band or an open-ended cord like a guitar string. By "little" I really do mean little. A typical estimate is 10^{-34} meter, although one could be as small as the Planck scale of 10^{-35} meter. To such a string, an atom is as big as the entire observable universe is to us.

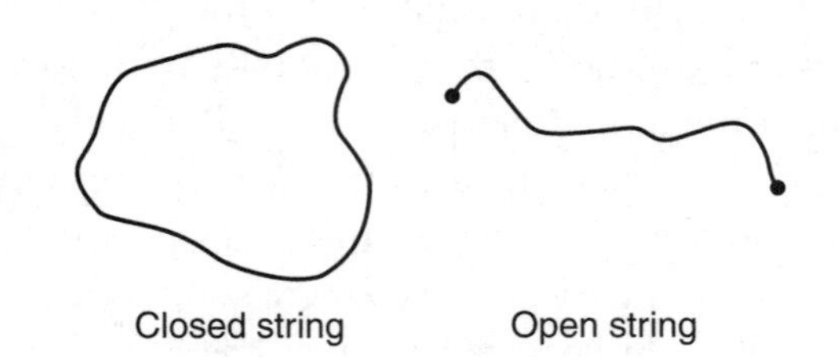

Closed strings form a loop, like a rubber band, whereas open strings are open-ended lengths, like a guitar string.

Driven by quantum effects, strings are always doing something; they never sit still. They can move through space; they can wriggle; and they can wrap around things like an anaconda (a very small anaconda). Different vibrational patterns correspond to different types of particles. Strings can't vibrate just any old way. Like musical instruments, they can play a series of fundamental notes and their overtones. For an open string, like a guitar string, a whole number of vibrations must fit along its length. For a closed string, a whole number of vibrations must fit around the circumference (which makes them more like a bell than a guitar). These restrictions organically explain the limited number of particle types. The fundamental notes of strings correspond to the particles of the Standard Model and the overtones to a heavier breed of particles.

A string can stop vibrating in one way and start vibrating in another, like a guitar string playing a new note. In this way a particle can metamorphose from one type to another—an electron, say, to a quark, or even a particle of matter to a particle of force. Two strings exert a force on one another by interchanging a third string serving as a middleman. To create the middleman, one of the strings undergoes a process a bit like cell division in the human body. It pinches off and creates a new loop, which flutters through space and gets absorbed by the other string.

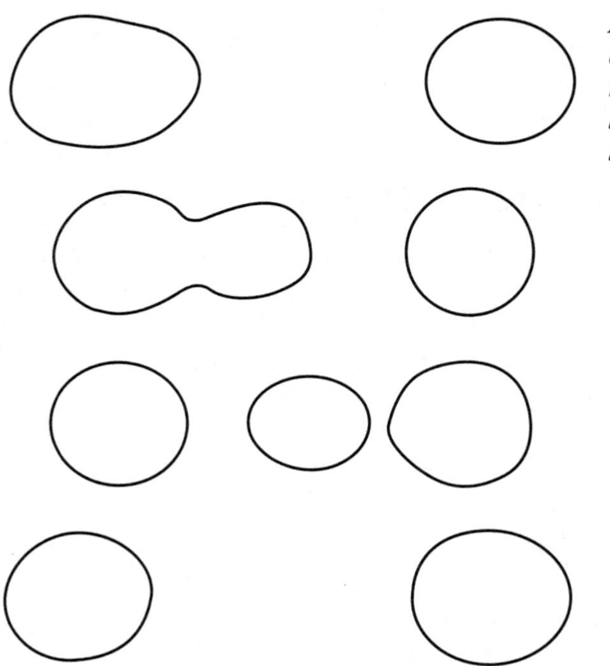

A part of a string can pinch off, creating a new string—hence a new particle—that another particle can then absorb.

Strings themselves determine how strong these forces are. One of the ways they vibrate is to pulse in and out like a tiny heartbeat. This vigor of this pulsation, called the *dilaton*, governs the gregariousness of strings and, in turn, the strength of gravity, electromagnetism, and the nuclear forces. The strength of these forces, which physicists often take as a fixed property of the universe, is the upshot of string dynamics.

def•i•ni•tion

The **dilaton** is a type of string vibration that governs the strength of string interactions and therefore of all the forces of nature.

The Inner Life of Strings

Attributing particle properties to the internal machinations of strings is easier said than done because we run into a potential contradiction with special relativity. It takes time for influences to cross a string. If we wiggle one end, the other doesn't respond immediately; the effect must ripple down the length of the string. Consequently, what happens to the string could become ambiguous.

Think back to the train examples in Chapter 3. One person sees two beams of light hit both ends of the train car at the same time, while another person sees the beams hit one after the other. For the train car, this poses no contradiction because the train car is not really a single unit but a collection of parts. What happens at one end is divorced from what happens at the other. For a string, which *is* supposed be as a single unit, we might get odd reversals of cause and effect. Some people might see the string respond even before we start wiggling it, which is tantamount to time travel.

What's remarkable about strings is that they can behave consistently in spite of it all. People may disagree on the position of events, but they agree on the timing, so a string acts as a unit after all. Because of the way events are defined for strings, the spread in position does not imply a spread in time. The position ambiguity is a positive benefit of the theory. Whereas the Standard Model crams an entire particle interaction into one infinitesimal point, string theory gives particles some breathing room to do their thing.

To stay consistent, strings must meet certain criteria, which are so restrictive that it's amazing strings can satisfy them at all. The criteria underpin the predictions of string theory. In this way, string theory takes one of the potential show-stoppers of quantum gravity and adopts it as an organizing principle.

The key is that strings have both an inner and an outer space. If you view a moving string in spacetime, it traces out either a tube (if it's a closed string) or a ribbon (if it's an open one). To specify a location on this surface, you need two numbers, like latitude and longitude on Earth's surface. For the string, one number gives the position along the string, the other gives the internal time for the string. All the real action takes place on this surface, known as the *worldsheet*—the inner space. The processes are then translated into events occurring in ordinary spacetime—the outer space. In string theory, ordinary spacetime is really a secondary concept.

def•i•ni•tion

The **worldsheet** is the two-dimensional surface that strings trace out as they move through spacetime.

What connects inner to outer is that the string arranges itself to minimize its area in spacetime. In other words, strings do the least amount of work they can get away with—many of us could sympathize.

Because strings are featureless, nothing singles out any particular location along them as special. Strings have no built-in concept of distance. The way we specify locations in their inner level is completely arbitrary. You could use a wooden ruler with regular markings; I could use a swizzle stick with randomly spaced scratches, and we'd both make the same predictions for how the string acts. If I were in a mischievous mood, I could even measure everything backward, so that when distance doubles for you, it halves for me. There's no objective distinction between short and long distance. For this reason, string theory doesn't get into any of the troubles with infinitesimal distances that ordinary quantum theory does.

Gravitating to Strings

Among the criteria governing strings' behavior, one stands out as the reason so many physicists are wrapped up in string theory. A closed string can vibrate by expanding in one direction and squeezing in the perpendicular direction, then squeezing in the first direction and expanding in the second, back and forth, over and over. And that's just what gravitational waves do (see Chapter 11)! So a string vibrating in this way must be acting as a graviton. Firing off gravitons is how particles exert the force of gravity on one another.

When gravity kicks in, ensuring consistency becomes even trickier than usual. The string vibrates and moves in a curved spacetime, and it takes an additional constraint to maintain order. This constraint turns out to be none other than the general theory of relativity. Whereas standard quantum theory buckles if you try to force-fit general relativity in, string theory unravels if you leave it out. If you had never heard of general relativity, you could deduce it by studying strings. No other proposed unified theory of physics can claim such a feat.

All Tangled Up

Are the strings little cords connecting particles? It's tempting to think of gravity as a bungee cord extending from the center of Earth to a satellite, holding it in orbit. It's tempting, but wrong. A vibrating loop behaves like a graviton, and this graviton then flies across space and does its thing. In other words, forces are still transmitted by particles, as in the Standard Model. It's just that the particles themselves are strings vibrating in a certain way.

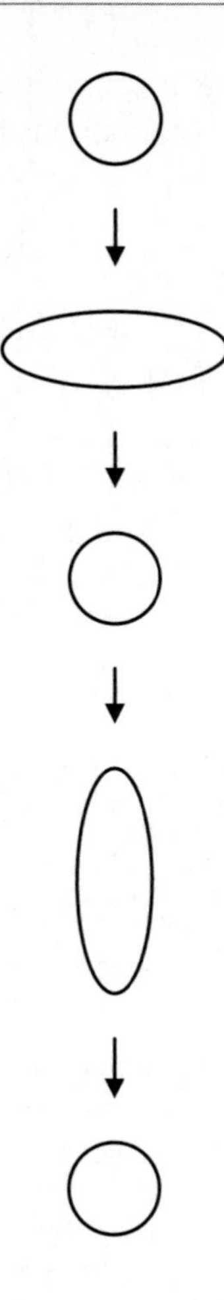

A closed loop can vibrate in just the way a graviton does (see Chapter 11).

What's more, the fact that strings extend over space solves the paradox of the graviton (see Chapter 11). Although the stringy graviton loops can split into new loops, which can themselves break up, and so on, this chain reaction eventually comes to an end when the loops are too small to divide any further, keeping gravity from spiraling out of control.

One downside is that, like the troublesome graviton theory, string theory doesn't treat spacetime as fully malleable. Gravitons transmit the force of gravity between bodies, but don't resculpt the spacetime around them. The way the theory handles the malleability of spacetime is indirect. It allows spacetime to take on different shapes and provides a set of rules for relating one shape to another. Ideally, though, a quantum theory of gravity should describe how the shapes morph. String theory provides a series of snapshots, but ultimately theorists would like a movie.

Brane Bogglers

Strings promise to explain all the known particles and forces of nature. For this purpose, they're ideal. One-dimensional objects are just big enough to solve the riddles of particles yet are not so big that they threaten to behave inconsistently. But why stop at the known particles and forces? What unknown things might lie out there? Physicists have found the theory also predicts novel objects from two-dimensional membranes that stretch across space like drumheads to lower- and higher-dimensional versions known as *branes*—a word that not only generalizes the concept of membrane but also allows for lots of puns and wordplay, which mischievous physicists have taken full advantage of ("pea brane," "brane scan").

> **def•i•ni•tion**
>
> A **brane** is a two-dimensional membrane or analogous object in lower or higher dimensions.

A 0-brane is a point, a 1-brane is a type of string, a 2-brane is a sheet, and a 3-brane is a solid body. The higher-dimensional branes are heavier, so they take more energy to create and are proportionately rarer. Branes exert forces on one another, buzz around like molecules in a gas, and orbit one another like planets and moons. You can even have "antibranes" that act like antimatter, capable of destroying any ordinary branes they encounter.

One type of brane plays a special role, the *D-branes.* The D comes from "Dirichlet," which refers to the fixing of the end points to a certain location and is named after a nineteenth-century German mathematician. D-branes stick to the loose ends of open strings like a sheet of flypaper. An open string can connect its two ends to one D-brane or span two different D-branes. The string is then pinned down like a guitar string, limiting how it can vibrate. Our entire universe may be glued to D-branes, a concept that I'll flesh out in Chapter 14.

> **def•i•ni•tion**
>
> A **D-brane** is a special type of brane that open-ended strings attach themselves to.

Theorists are trying to decide between two ways of thinking about branes. Are branes secondary, maybe legions of strings wriggling in unison, or are they primary, indivisible building blocks in their own right? If the latter is true, 0-branes could play a central role. These branes, which are really just an exotic type of particle, flit around and can cause strings to wriggle. It's a case of "the particle is dead, long live the particle." Physicists took to strings to get around the contradictions posed by quantum particles, yet many of them are coming back around to a particle-based theory. Unlike the particles of the Standard Model, 0-branes don't arise from undulations in a quantum field, so they avoid the contradictions of earlier efforts.

D-branes are sheets that open-ended strings stick to.

(Copyright 2004 Steuard Jensen, Joint Science Department of the Claremont Colleges)

String theory has a strong claim to be a candidate for "theory of everything," but this label is both its glory and its bane. It's the reason people get so excited about the theory and also the reason it has gone through cycles of boom and bust. People's prejudices about what a final theory should look like have colored their reaction to strings. A little expectations-management would serve string theory well. The theory does well in the "everything" department; it's the "theory" part that still needs work. The theory has introduced physicists to a wide range of new concepts that will survive whatever ultimately becomes of the theory itself.

The Least You Need to Know

- String theory once languished in the backwaters of physics but has slowly become the leading approach to a quantum theory of gravity.
- According to the theory, nature's basic building blocks are strings.
- Each elementary particle is a string vibrating in a certain way.
- Strings behave consistently only if a large number of criteria are met, which gives the theory its conceptual power.
- Because strings have an intrinsic size, they tame the infinities that plague ordinary particle theories.
- Although the theory singles out strings as special, it has other important ingredients called branes.

Chapter 13

Playing a Different Tune

In This Chapter

- The main alternatives to string theory
- Piecing space together
- Space without space
- Turning physics on its head

String theorists sometimes say their theory is "the only candidate," "the only viable attempt," or "the only game in town." This isn't true, and not surprisingly, such statements annoy those working on alternative theories. Although these alternatives are not as well-developed as string theory, each introduces provocative ideas.

What Else Is out There

String theory comes out of a particular school of physics—"particular" in both senses of the word. The theory is rooted in particle physics and inherits its culture and temperament. But particle physics is hardly the only physics there is. Several alternatives to string theory are rooted in relativity theory, a subdiscipline with a very different style. Relativists are a smaller, tighter-knit community. They consider geometry to be fundamental rather than particles. Physicists from a third subdiscipline, those who study fluids and solids, have also put forward a candidate quantum theory of gravity. They tend to think that what happens at the fundamental level is irrelevant and what truly matters are the laws that connect the micro to the macro.

In the tradition of Albert Einstein, relativists love confronting deep philosophical questions. String theorists, in the tradition of those who developed the Standard Model, tend to shoot first and ask conceptual questions later. The fluid-and-solid people hew closely to experimental analogies. And each group can point to historical precedents for the success of its approach.

This scientific and sociological diversity carries over into the theories themselves. The different schools have different thoughts to what the mental interlopers—the extraneous features of current theories that block their unification—might be. All claim to be the "radically conservative" option, the one that minimizes the changes that have to be made to current theories. Although there's some bad blood among the different schools (see Chapter 22), it makes sense to attack such an entrenched problem on multiple fronts.

Loop Quantum Gravity

The leading alternative to string theory goes by the name of loop quantum gravity. Like string theory, loop gravity went through several near-death experiences in its formative years. An early incarnation hit a brick wall in the 1970s and was left for dead. It sprang back to life in the mid-1980s. The key figures in this resuscitation were Lee Smolin of the Perimeter Institute near Toronto, Carlo Rovelli of the University of the Mediterranean in Marseille, and Abhay Ashketar of Pennsylvania State University as they drew on earlier ideas of Roger Penrose of Oxford University.

Loop gravity seeks to quantize general relativity rather than derive it from particle interactions. It takes the principles of relativity as its starting point. This approach is both less and more ambitious than that of string theory. On the one hand, loop gravity doesn't set out to be a theory of everything, just of gravity (not that this isn't ambitious enough). On the other hand, it dives straight into the conceptual deep end of quantum gravity: namely, the problem of concocting a theory when not even the structure of spacetime can be assumed in advance.

Loop-d-Loop

If someone asks you for directions to a bar in Manhattan, you can answer in two ways. Either give the street intersection, which works well if the person wants to drive or take a cab, or give the subway line and station. Similarly, you can describe space in two equivalent ways: as a grid of points or as a pattern of lines. The original equations of general relativity use points, but in the mid-1980s, physicists rewrote these equations in terms of lines. These lines correspond to the lines of gravitational force, much like the magnetic lines around a bar magnet, which indicate the direction that a compass needle points. These lines can loop back on themselves, and loops can wind through other loops like a tangle of subway lines. They are the "loops" of loop quantum gravity.

Loops are more abstract than points, but they neatly capture an essential principle of Einstein's theory. If two people observe the same region of space, they may think it has a different shape, but this difference is not real; it's just an artifact of their two vantage points. In terms of loops, a mere shift of vantage point corresponds to sliding the loops around without altering which loop connects to which other loops. It's like one of those tavern puzzles, which consist of interlocking metal pieces that you're supposed to disentangle. If you jiggle the puzzle pieces a little bit, you get nowhere in solving the puzzle—you have not really changed the puzzle conditions. Only if you slide the pieces in just the right way can you pull them apart.

The flexibility of loops makes them easier to work with mathematically than the traditional continuum of points. Physicists can turn the loopy version of relativity into a quantum theory of gravity using the same procedure that turned the prequantum theory of electromagnetism into the quantum theory of electromagnetism. The same procedure chokes when applied to the continuum version of relativity (see Chapter 11).

Atoms of Space

Once you've got the quantum version of relativity, you can translate from the loops back into ordinary space, and the first thing you notice is that the points have all gone away. Instead, space comes in discrete chunks—"atoms" of space. These atoms (I guess we could call them "spatoms") represent the smallest possible units of volume: 1 cubic Planck length. The fact that they're not true points makes a difference. If space cannot be subdivided any more than a space atom, it limits the strength with which gravity acts on particles. So these atoms are loop gravity's answer to the paradox of the graviton.

The equivalent vantage points of different people correspond to shifting the arrangement of atoms without altering which connects to which. Real changes to space correspond to a reconfigured pattern of interconnections. In fact, loop theorists commonly forget about the space atoms themselves and think only of the web of their relationships, depicted on diagrams that look like connect-the-dots puzzles. Such a diagram, though drawn on a flat page of paper, can represent a multidimensional space. A nice regular grid might correspond to empty space, devoid of gravitational forces. An elaborate web of links might portray a region of intense gravitation.

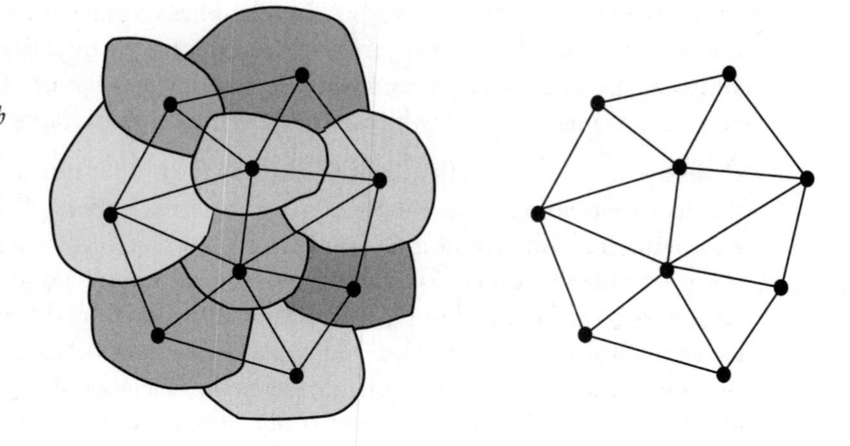

According to loop quantum gravity, a volume of space consists of discrete chunks—"atoms" of space (left), which can be represented as dots (right).

(Courtesy of Carlo Rovelli)

Despite the loop theorists' protestations about their limited ambitions, in the back of their minds, many of them always thought that a full theory of spacetime would automatically be a theory of particles, too—hence a theory of everything. And there are tentative indications they were right. The links between atoms can pass through and around one another, creating knots and braids. It takes three strands to make a braid that can't undo itself. Lo and behold, the various permutations of interweaving three strands match up with the particles of the Standard Model. Some braid patterns have a twist in each of their strands, others in only one or two strands; these could represent electrons and quarks, with their differing amounts of electric charge. To be sure, this interpretation of particles is still rather sketchy.

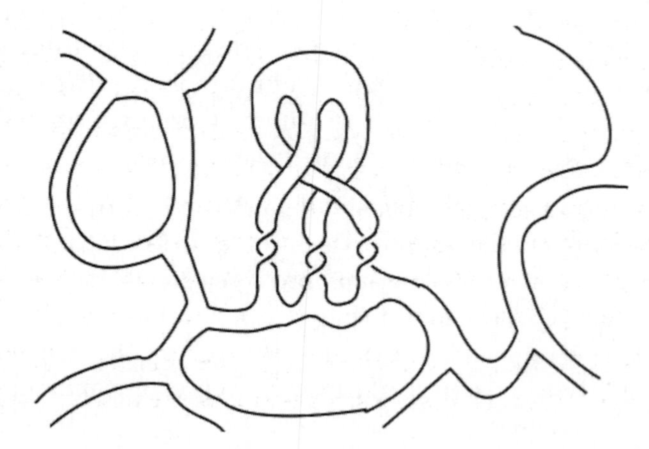

Braids in the interconnections of loop quantum gravity have all the hallmarks of particles (in this case, a positron or antielectron).

(Courtesy of Sundance Bilson-Thompson)

Where Do Loops Stand?

The great success of loop gravity is to describe space as a dynamical entity rather than a fixed framework. Many questions remain, however. Over short distances, the theory should reproduce the predictions of special relativity. Does it? Over long distances, the details of the individual space atoms and their interconnections should fade to insignificance, so that all you'll see is their averaged behavior. Does this behavior match general relativity? Theorists don't yet know.

You may have noticed that I've been using "space" rather than "spacetime." That's not a typo. The procedure for bootstrapping from Einstein's equations up to quantum theory dismantles spacetime into space and time. To put it back together, loop theorists extend their networks of interconnections using a set of mathematical principles that are not as well developed. Moreover, the dismantling leads to the problem of frozen time (see Chapter 7). That is, the time variable in the theory doesn't jibe with our common-sense notion of time.

String theorists by and large don't think too highly of loop quantum gravity, and the feeling is reciprocated. What string theorists like most about their theory and expect from other theories—ensuring that special relativity holds, deriving general relativity as an approximation, and predicting a rich assortment of phenomena—are precisely what loop theorists don't like. Some in each school, though, are struck by intriguing parallels between loops and strings. Both are extended objects that interconnect with one another and come in a certain minimum size. Might the two theories prove to be different trails up the same mountain?

"Buckyspace"

In quantum gravity, there almost seems to be an inverse relationship between the complexity of a theory and the complexity of its name. "String theory" is a simple name for an unsimple theory. "Causal dynamical triangulations" is a mouthful for what is really a very simple theory. All it does is approximate space with a scaffolding of triangles, like a higher-dimensional version of Buckminster Fuller's geodesic domes, so, to give this theory a simpler name, unofficially I call it "buckyspace." During the past couple of years, Renate Loll of Utrecht University, Jan Ambjørn of Copenhagen University, and Jerzy Jurkiewicz of Jagellonian University in Krakow, Poland, have fit this theory together.

The idea builds on an approach developed by Stephen Hawking, among others. Remember that quantum theory holds that what we observe emerges from a mixture of all possibilities (see Chapter 4). Applied to the universe, it suggests that the spacetime we observe emerges from a goulash of all possible spacetimes, and Einstein's equations tell us the proportions of each ingredient to throw in the pot.

To perform the calculations, we need to find a common language for describing all these shapes. That's where the triangles come in. Enough of them can approximate any shape at all, which is why computer animation uses them to render aliens and bombed-out landscapes. The word "causal" in the formal title of the theory refers to how the triangles have a built-in distinction between time and space and are stitched together in a way that forbids time machines. These attributes ensure that cause-effect relations remain consistent, which is one of the trickiest aspects of any quantum theory of gravity.

Proponents of this approach don't claim that spacetime is literally a scaffolding of triangles. It's just an approximate description. If you use sufficiently small triangles, the approximation should be indistinguishable from reality. Similarly, if you build your geodesic dome out of teeny-weeny triangles, no one will be able to tell it apart from a smooth sphere.

After throwing the mixture of triangle-spanned spaces into their mathematical pot, physicists stir it around and see whether our observed 4-D spacetime emerges. It does. What's more, spacetime takes on a very different shape on fine scales than it does on large ones (see Chapters 14 and 16), which keeps gravity from going haywire.

The technique is still in its infancy and it's somewhat of a "black box." It tells them what space does, but not why. Numerous issues remain open. Do the details of the building blocks introduce a subtle bias into the approximation? Can the buckyspace theory accommodate matter? Can it accommodate black holes and other exotic phenomena? The answers to all these questions are maybe. The technique has caught the eye of both string theorists and loop theorists, who see it meshing with their own efforts.

Domino Theory

Earlier, I talked about giving directions in terms of either a street grid or subway lines. There is, of course, a whole other paradigm: landmarks. Instead of "go north three-tenths of a mile, then east," you could say, "keep going until you see a big white rock under the an oak tree, then bear right." Some prefer the orientation-and-distance strategy; others the landmarks. Most proposed quantum theories of gravity describe spacetime in terms of abstract coordinates, but some use landmarks. These include the work of Rafael Sorkin and Fotini Markopoulou of the Perimeter Institute and Fay Dowker of Imperial College London.

For spacetime, the concept of "landmark" gets generalized to *event*, which combines a physical landmark with a time, as in "meet me under the big clock at four minutes of two." You can flip it around and say the event defines the position. Once you take events as primary, your entire perspective on spacetime changes. Instead of a continuum of points, what really matters is the web of events and the relationships between them. One event causes another, which causes another and another, in a never-ending domino effect. Physicists call it a *causal set.*

def•i•ni•tion

An **event,** to physicists, marks a specific position in space at a specific time. It is the basic unit of the spacetime continuum. A **causal set** is one possible family tree of these events: a network of events organized according to which caused which.

Physicists draw causal sets using connect-the-dots diagrams that look superficially like those in loop quantum gravity but have one gigantic difference. The loopy diagrams are the end point of a series of calculations involving loops and Einstein's equations. The causal sets are a starting point. To draw one, we do not even need to assume relativity. Proponents figure out the multitude of ways that events can be ordered and show that concepts such as distance, time, and gravitation emerge when we consider events en masse. If we apply quantum theory to the events, we find there are only certain ways to assign them positions that are consistent with all the relationships they have.

Quantum Leap

Could our universe be a giant computer? In essence, that's what the causal-set theories say. Space comes in discrete chunks like bytes in a computer memory; time marches forward in discrete increments like ticks of a computer clock; and events are related like logical operations in a computer program. A number of computer scientists, such as Seth Lloyd of M.I.T., Edward Fredkin of Carnegie Mellon University, and Stephen Wolfram of Wolfram Research, have explored this connection.

One appealing aspect of the approach is that it avoids the problem of time. It builds in a common-sense notion of change from the outset. Nor does the graviton problem ever arise; the inherent discreteness of events forestalls it. A big problem, though, is that there are countless ways to order a set of events, and only a tiny fraction look anything like our universe.

A Tipping Point?

Finally, a handful of physicists think the approaches to quantum gravity we just explored are badly misguided. They argue that the most important laws are those describing how parts assemble themselves into wholes. The laws of the parts themselves are beside the point. Maybe the parts don't obey any laws—total anarchy reigns. Holger Nielsen developed this idea of "anti-grand unification" in the 1970s, and other proponents include Grigori Volovik of the Helsinki University of Technology and Robert Laughlin of Stanford University.

def•i•ni•tion

Universality is the principle that very different substances can all have the same behavior under certain circumstances. In those situations, the details of their composition cease to matter.

These researchers draw their inspiration from the behavior of fluids and solids. When water freezes, the H_2O molecules do not change; only their collective behavior does. Freely swimming molecules slow down until they reach a tipping point when all latch on to their neighbors and lock one another in place, forming an ice crystal or glassy material. Other substances do the same when they freeze. This affinity among all substances is known as *universality*.

As it happens, fundamental theories bear an uncanny resemblance to the laws of fluids and solids. In Chapter 8, I compared black holes to water flowing down a drain. This is not just a helpful image. The equations of sound propagation through a moving fluid are a dead ringer for the propagation of light through spacetime. Does universality extend to spacetime itself?

That opens up a whole new interpretation of relativity theory. The theory might describe the collective behavior of nature's basic constituents rather than the constituents themselves. What those constituents are hardly matters. Even if a theorist hazards a guess, an experimenter could never verify it. A million guesses could produce the same outcome. And if the details are immaterial, they might as well be completely random. The tree of physics could be rooted in anarchy.

"Random" is the word most other physicists would use to describe this approach to quantum gravity. To them, the idea that nature becomes anarchic at its deepest levels runs contrary to the observed trend of increasing order. The strategy of reductionism—of finding explanations in terms of ever-smaller components—has proven so successful that physicists are reluctant to abandon it just yet. Universality does not always hold, so experiments can, in fact, distinguish among theories. As for the resemblance of fluids and relativity theory, it could reflect commonplace mathematical properties, so it shouldn't be taken literally.

And yet string theorists find themselves rediscovering many of the same ideas. In certain settings, general relativity does seem to emerge in the act of aggregating particle behavior (see Chapter 18). String theorists also find phase transitions and collective behavior popping up all over the place. Related ideas have made their way into loop gravity as well. These similarities are yet another reason that researchers need to tackle quantum gravity on all sides. Each researcher sees things the others might have glossed over.

The Least You Need to Know

- String theory is not the only candidate for a quantum theory of gravity.
- The main alternative, loop quantum gravity, breaks a volume down into "atoms" of space.
- A newer technique approximates spacetime as "buckyspace," like a multidimensional geodesic dome.
- Another theory describes spacetime in terms of events rather than points.
- One radical proposal changes the focus from fundamental laws to the laws of collective behavior.

Part 5 The Big Ideas

Even in their current incomplete state, string theory and other such theories lead to a huge expansion in our conception of reality: higher dimensions of space, parallel universes, novel behavior on the finest scales, new interrelationships among types of matter and forces, and the idea that space and time are composed of something spaceless and timeless. It's an intellectually fertile time when people are developing new ways of thinking about some of the deepest questions possible. We'll step through the ideas and find out what the diverse theories have to say about them.

Chapter 14

Extra Dimensions

In This Chapter

- Visualizing the unvisualizable
- Ways to feel extra dimensions
- The funky shape of higher space
- Why three is special

String theory works best when strings have plenty of elbow room. Ordinary space, with its three dimensions of length, width, and height, cramps their style. A total of 10 space dimensions is ideal. We can't see or roam around the extra dimensions, either because they're too small for us to fit into or because our 3-D world holds us captive. But if they exist, we might be able to perceive them indirectly.

Headed in a New Direction

Is there a galaxy right next to you? It might be just an eyelash's width away, yet you can't see it, feel it, or duck over to it for some peace and calm. To get there, you'd have to travel in a fourth dimension of space, one that is perpendicular to the usual three of length, width, and height. You can't even point to it. It's all around you, everywhere and nowhere at once.

Blown minds are an occupational hazard in string theory, never more so than when talking about one of the theory's most distinctive aspects: extra dimensions of space. I'm not referring to time, which can be thought of as a fourth dimension (though a somewhat different one than the usual three), but new honest-to-goodness dimensions of space. If you could avail yourself of the extra freedom of motion they provide, your friends would cower in fear of your superpowers. You could reach into a locked safe by bending your arm around through an extra dimension. You could turn right-handed baseball pitchers into southpaws by flipping them over within the extra dimension. You could untie a tightly knotted rope by sliding it apart in the extra dimension. To adapt a saying of novelist Arthur C. Clarke, any sufficiently advanced notion of space is indistinguishable from magic.

String theory suggests that space has a total of nine dimensions—six more than the ones we see. M-theory, which underlies string theory, adds yet one more. Including time, that makes a total of 11. The reason for them goes back to the question of consistency (see Chapter 12). If strings lived in plain 3-D space, they'd violate Einstein's special theory of relativity. Some researchers have explored other ways, besides piling on dimensions, to keep strings from freaking out, but most think that the extra dimensions are unavoidable.

Quantum Leap

Why would space have 10 dimensions rather than, say, 42? In 10-D space, string theory is able to marry all the particle dynasties together. The electron pairs off with a sibling of the photon, which in turn finds itself related to more exotic particles. The end result is one big happy family. If there were fewer than 10 dimensions, particles would split into isolated clans. If there were more than 10, the family would admit some black sheep—additional particles that refused to interact with the others. Either all particles must interact or none can, so the presence of a few antisocial particles would force all the other particles to cease interacting as well.

All this is still speculative. Not only have experimenters yet to detect unequivocal signs of extra dimensions, but also proponents of alternatives to string theory see no need to go beyond 3-D space. Nonetheless, it's fun to explore the concept of extra dimensions. And if physics teaches us anything, it's that many seemingly obvious features of the world are artifacts of the peculiar position we occupy within it. Someone who'd never left the Himalayas might laugh if you told them of the exotic place known as Nebraska, where two dimensions are quite enough to describe your position. Likewise, the fact that space looks 3-D might be an artifact of our position within it.

From Flatland to Hyperspace

Let's face it: visualizing extra dimensions is impossible. That's what makes them so much fun. Once you accept that evolution did not equip your brain with the capacity to conceive of a fourth dimension, let alone a fifth or higher, you free yourself to explore them without any pressure—never quite grasping what it would mean to ramble off the beaten path of ordinary space, but having a series of "aha" moments as you reflect on them from different angles.

The tried-and-true technique, going back to Edwin A. Abbott's classic nineteenth-century novel *Flatland*, is to imagine what our 3-D world would look like to a 2-D creature. From there, we can step up to imagining how a higher-dimensional space would look to us.

So let's start with zero dimensions, a mere point. By piecing together points, we create a 1-D line. By lining up lines, we create a 2-D square. By stacking squares, we create a 3-D cube. The dimension of the geometric object indicates how many numbers we need to describe position within that object. A point requires none. A line requires one, indicating where we are along its length. A square requires two, for length and width. A cube adds height.

0 dimensions (point)

1 dimension (line)

2 dimensions (square)

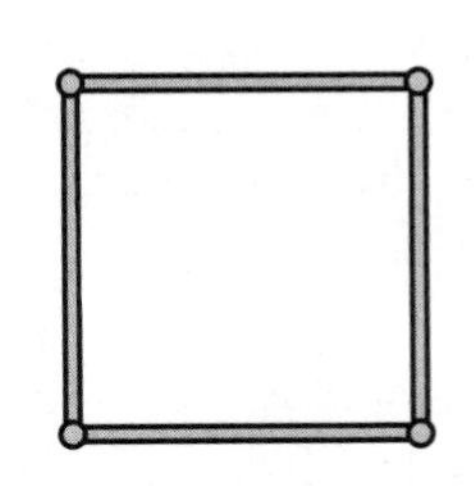

3 dimensions (cube)

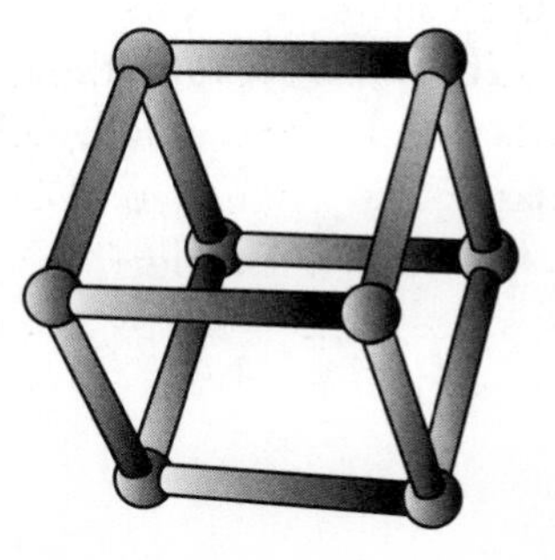

Points lead to lines; lines lead to squares; and squares lead to cubes. What's next?

Why stop there? By connecting cubes, we create a 4-D *hypercube*, also called a tesseract. To length, width, and height, the hypercube adds some indefinable direction perpendicular to all three. A square has four sides (lines); a cube has six sides (its square faces); and a hypercube has eight sides (cubes). If we double the width of a square, we increase its total size (area) fourfold. If we double the width of a cube, we increase its total size (volume) eightfold. If we do the same to a hypercube, we increase its total size (hypervolume) 16-fold.

def•i•ni•tion

A **hypercube** is a 4-D cube; a **pentacube** is a 5-D one; and a **hexacube** is a 6-D one.

A **hypersphere** is a 4-D sphere, and a **hyperpyramid** is a 4-D pyramid.

Those brave enough to venture into five dimensions find *pentacubes*, and if you dare go into six, *hexacubes* await you. In principle, it never ends. Other geometric figures follow the same progression. As you bump up the number of dimensions, a circle becomes a sphere, then a *hypersphere*. A triangle becomes a tetrahedron (a type of pyramid), then a pentachoron (a *hyperpyramid*). The area, volume, and hypervolume follow the same increasing trend for them as for the cubes.

Imagine rotating a 3-D, wire-frame cube. In the perspective view shown here, it looks like a small square (the back of the cube) inside a bigger one (the front). Surrounding the small square are four trapezoids, which are actually squares (the faces of the cube) distorted by the perspective. Give your eyes a few seconds to catch on. If you need, ask a toddler if you can borrow a toy wooden block. If you rotate the cube counterclockwise as seen from above the cube, the rear face ends up on your left side, and the squares get bent.

A flattened cube, rotating clockwise as seen from above.

(Courtesy of Drew Olbrich)

Now step up from a cube to a hypercube. When it is "flattened" from four to three dimensions, it looks like a small cube inside a bigger one. The small, dark cube is the back of the hypercube. It's small because it's farther away in this perspective view. The six pyramidlike shapes around it are actually cubes (the "faces" of the hypercube) distorted by the perspective.

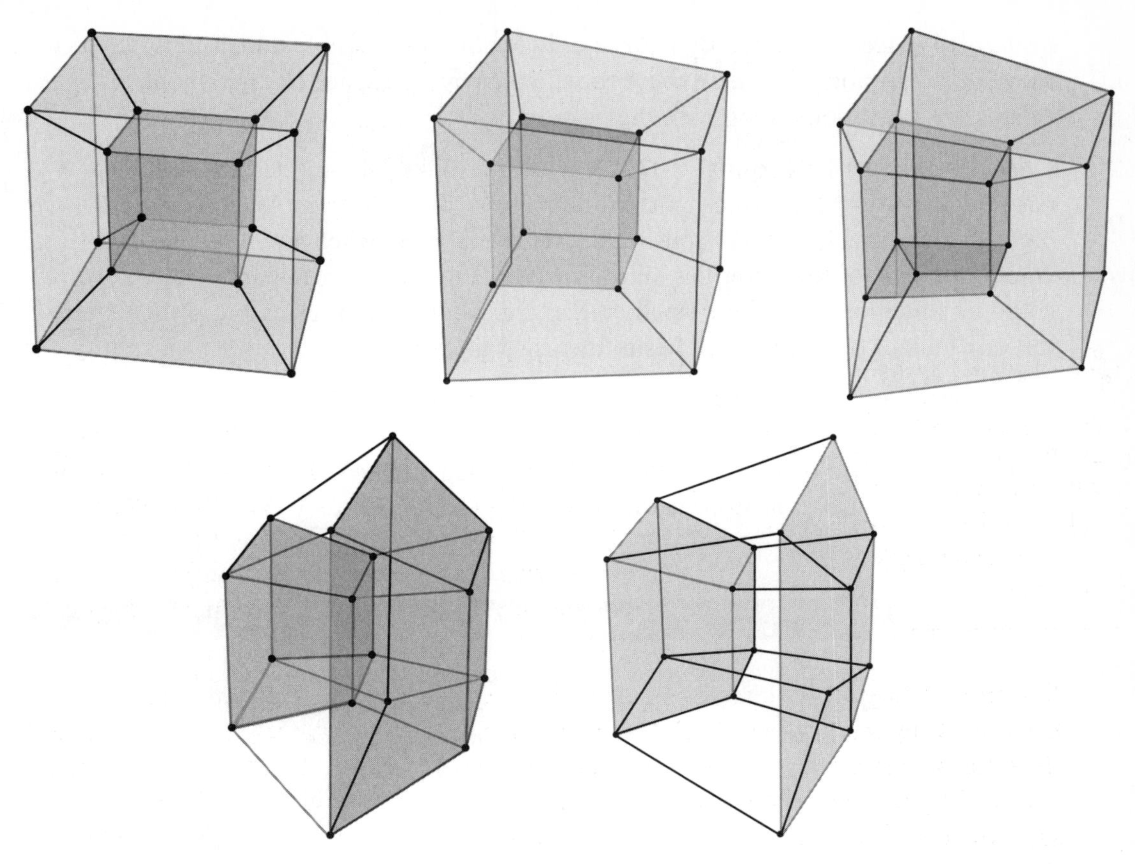

A flattened hypercube, rotating in the fourth dimension.

Dozens of websites have other figures that help. I list a few in Appendix B, or you can just do a search for "hypercube."

Running Out of Space

Why oh why are we so unlucky? How did we get stuck in only three dimensions, unable to behold the full glory of space? Broadly speaking, string theorists have identified two possible reasons. Either we're too big to fit into the extra dimensions, or our very existence depends on our staying put.

Most of us think of space as novelist Douglas Adams memorably described it: "Space is big. Really big. You just won't believe how vastly hugely mind-bogglingly big it is. I mean, you may think it's a long way down the road to the chemist's, but that's just peanuts to space." Yet that may be true only for the three visible dimensions. If you move in a direction other than these three, you may find space claustrophobic. Maybe Adams was on to something with the peanuts.

Consider a flat parking lot. It's 2-D. Or at least it looks 2-D. In fact, the pavement has bumps and cracks, giving it a third dimension: that of depth. As long as these blemishes are small, your car rolls right over them. Even when a puddle fills the cracks, the bulk of the water lies on the surface. The extra dimensions of the universe could be something like these small cracks. Our bodies might extend into them, but the vast bulk of our bodies still lies in the usual three dimensions.

So how small is "small"? Physicists aren't sure, but can estimate their maximum possible size. Particle accelerators have zoomed in on details as fine as 10^{-18} meter, about a thousandth the size of atomic nuclei, without seeing any new dimensions. If the size of the dimensions is all that stops us from entering them, then they can't be any bigger than that.

Quantum Leap

Suppose you shrank yourself to minuscule size, so that you fit into the extra dimensions. How fun would that be! If you took a walk through those dimensions, you could circle around the entire breadth of space and return to your starting point, just as if you took a balloon ride around Earth or played the video game Asteroids and zoomed off the right side of the screen and came back in the left. Or you might reach a true edge of space, a hard wall you'd bounce off. Light would bounce off it, too, so it'd look like a mirror.

Alternatively, the visible dimensions, rather than the extra ones, may be the sticking point. Maybe they just can't bear to let us go. Consider a fridge magnet. It can slide around in only two dimensions, up/down and left/right. The force of magnetism holds it to the fridge and prevents it from exploring the third dimension. Similarly, our bodies may be stuck to our three dimensions. The extra-dimensional space could be "vastly hugely mind-bogglingly big" after all. But we can't break free from 3-D space to venture into it.

def•i•ni•tion

A **braneworld** is a D-brane or set of D-branes that hosts a collection of particles—a self-contained universe.

String theory has a natural way of explaining this stickiness. If the particles in our bodies consist of open strings, their ends must be planted in a three-dimensional D-brane. Our observable universe could be such a brane, a so-called *braneworld*, floating within the full 10-D space like a leaf in the wind, with us as caterpillars clinging on them for dear life.

If we try to pull an open string out of the brane, the string gets longer, like a bungee cord, but the endpoints remain attached to the brane. Not even light can break free. If you're a cynic, you can think of the D-brane as our prison. If you're the hopeful sort, you can think of it as our cozy niche in an otherwise hostile universe.

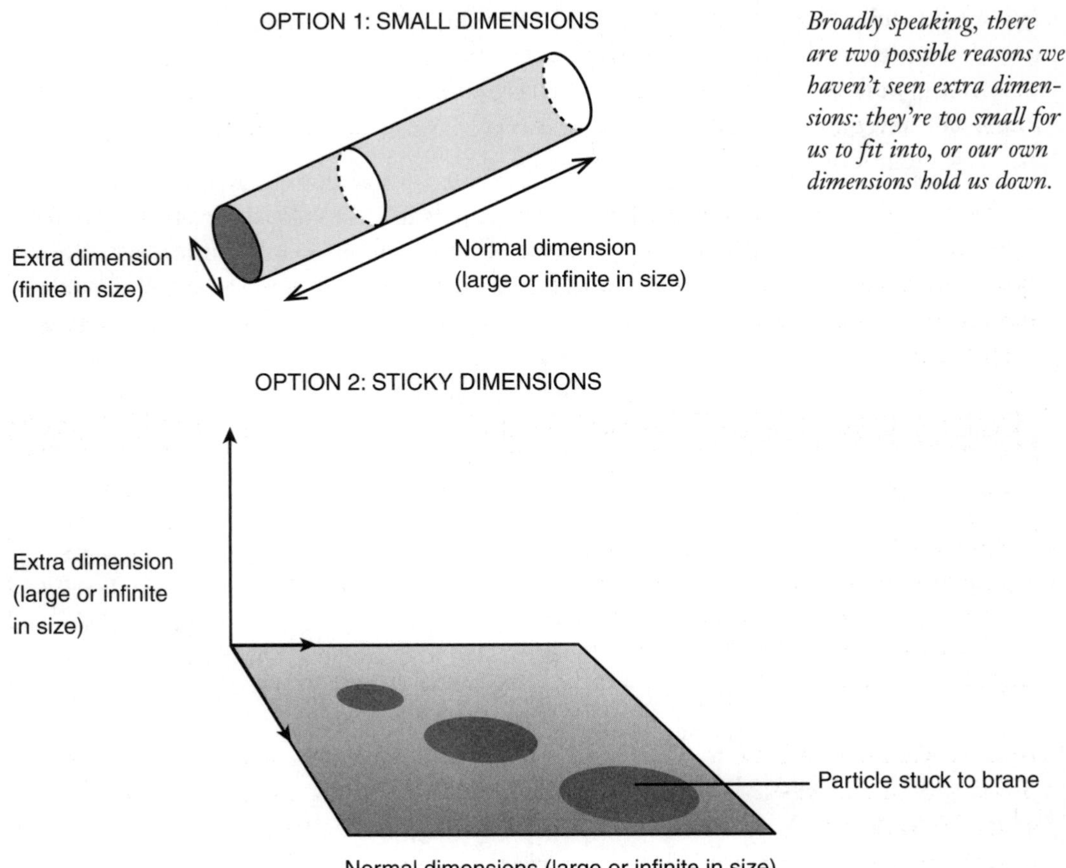

Broadly speaking, there are two possible reasons we haven't seen extra dimensions: they're too small for us to fit into, or our own dimensions hold us down.

Escaping the Shackles

Recognizing one's limitations is the first step to overcoming them. Although we may not be able to see the extra dimensions, we might see their effects if we know how to look. It depends on which of the two explanations we just explored applies.

Dimensional Shadow Puppets

In the small-dimension scenario, the only thing preventing us from venturing into the extra dimensions is our size. But a particle doesn't have to fit entirely into the dimension to feel it. Going back to the parking lot, if we pay close attention, we can tell when we drive over a crack. Our car rattles slightly. The smaller our tires, the worse it rattles. A shopping cart, pushed across a supposedly smooth pavement, can shake enough to crack the eggs. Similarly, a small particle could rattle in a small extra dimension of space. In effect, part of the particle dips into the extra dimension.

If you didn't know about pavement cracks and you saw a shopping cart rattle, you might think that something was wrong with the cart itself. Likewise, a rattling particle seems to be doing something, but we can't make out what it is, so we interpret it as some internal property. One of the most remarkable things about extra dimensions—and the reason they have such an appeal for physicists—is that this property has all the attributes of electric charge. If the extra dimension has the right shape, gravity acting within it looks like electromagnetism. Light, according to this idea, is the result of particles' dimensional derring-do. So although we may not see the extra dimensions, they may be the reason we can see at all. The strong and weak nuclear forces, too, could be the shadows of an extra-dimensional world.

In fact, extra dimensions might solve the hierarchy problem (see Chapter 7). As we dial up the energy, the forces of nature vary in strength at a very gradual rate. It takes a vast increase in energy for the forces to equalize. Extra dimensions might narrow this perplexing gap in energy. The particles' rattling provides new ways for forces to make themselves felt, and the forces vary in strength at a faster rate than they would in 3-D. They could equalize at a much lower energy than expected.

Footloose Gravity

In the sticky-dimension scenario, particles don't rattle, but they might have other ways to poke into the extra-dimensional space. If we pump enough energy into a string stuck to a brane, we can stretch it until it snaps in two. The result is a shorter version of the original string plus a closed loop—that is, a graviton. The graviton, having no ends to tie down, slips the surly bonds of D-branes and wanders off into the extra dimensions. It could be our blind man's cane, an implement to poke into a world we can't otherwise see.

Quantum Leap

Careful observations could look for signs that gravitons roam a higher-dimensional space. Highly energetic events such as stellar explosions would be a good place to watch for gravitons flying the coop. Black holes, on account of their intense gravity, could be sensitive to what happens in extra dimensions, too. And if our observable universe is only one braneworld among many, those other braneworlds could pull on ours gravitationally. Unable to tell where this force is coming from, we'd perceive it as having no direction at all. Such a force might account for various cosmological phenomena (see Chapter 20).

Gravitons' escape artistry would explain a big mystery about gravity: why it's so much weaker than the other forces of nature. Intrinsically, maybe it *isn't* weaker. It might just look that way because, unlike other forces, it rushes into the extra dimensions and gets diluted. If so, we may have extra dimensions to thank for the fact that Earth's gravity pulls on us gently rather than crushing us like a bug.

The tricky thing is that gravitons must be free to wander but not too free. If they all leaked out of our dimensions, we'd hardly feel gravity at all. The simplest way to plug the leak is for the extra dimensions to be modest in size (though still bigger than the small-dimension scenario). Gravitons can then saturate them like water in the pavement cracks. "Modest" means smaller than about 0.1 millimeter, the limit of current measurements of gravity.

A second way to plug the leak is for the extra-dimensional space to have some built-in energy, a variant of the cosmological constant (see Chapters 7 and 9). This energy curves space so strongly that gravitons find it hard to climb out of our brane. In effect, our 3-D universe sits at the bottom of a steep-walled canyon, and gravitons don't leak out because they can't scale the walls.

A third possibility is that other particles living on the brane restrict gravitons' freedom. If a graviton attempts to leave the brane, it skews the brane's particles, creating a gravitational force that draws it back in. If a graviton tries to enter the brane from the outside, it again causes an imbalance of particles, this time repelling the intruder.

This third option applies only to gravitons within a certain range of wavelengths. If the wavelength is too short, the gravitons oscillate too rapidly for the particles to respond, so they enter and exit at will. If it's too long, the gravitons are too pitifully weak to provoke a response, so they slip under the radar screen and again propagate as they please. This long-distance behavior might explain why cosmic expansion is accelerating (see Chapter 20).

On the Funky Side

Using extra dimensions, string theorists have tried to explain the specific pattern of particles and forces found in the Standard Model and its simplest extensions. In the small-dimension scenario, reproducing the Standard Model requires the extra dimensions to have a very specific shape. They must crumple up into a funky thing known as a Calabi-Yau shape, named after the two mathematicians who had worked out the details long before physicists took an interest. Calabi-Yaus come in progressively more complex varieties depending on how many extra dimensions there are. With two dimensions, things would be blissfully simple. The sole 2-D Calabi-Yau space is a torus—a doughnut-shaped space with one hole in the middle. Higher-dimensional versions are unimaginably convoluted pretzels with multiple holes.

Two-dimensional Calabi-Yau shape, which is a torus.

Calabi-Yaus and related shapes make sense of several of the idiosyncrasies of the Standard Model. For instance, their doughnutlike holes could account for why particles come in distinct generations. Each generation involves the same vibrational patterns, the only difference being how strings wrap through the holes. The shape of extra dimensions could also shed light on the mirror-asymmetry of the weak nuclear force. A mirror reflection in 3-D is a rotation in 4-D. If a baseball coach could somehow lift his star pitcher into a fourth dimension and flip him over, what used to be his right hand would become his left. The distinction between righties and lefties would evaporate. Sadly, we can't do that. Whatever hinders our motion through the extra dimensions may break the symmetry of left and right.

Four-dimensional Calabi-Yau space, sliced and projected into a 2-D image.

(Copyright 2005 Greg Egan)

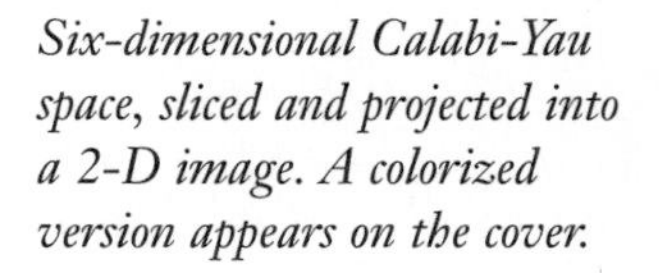

Six-dimensional Calabi-Yau space, sliced and projected into a 2-D image. A colorized version appears on the cover.

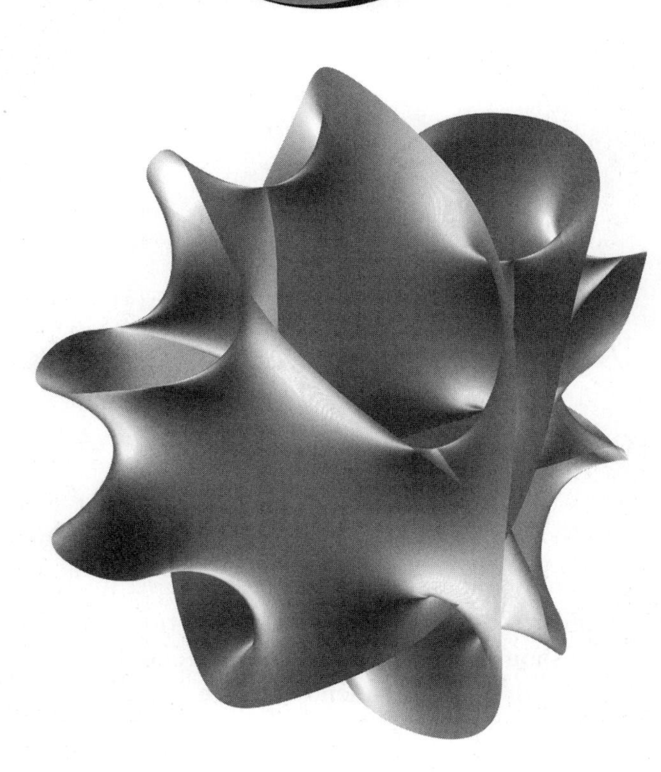

(Generated using software by Andrew Hanson and Jeff Bryant)

Unfortunately, string theorists have trouble going beyond these generalities. Performing calculations on complex spaces is, well, complex. Mathematicians don't even know how many 6-D Calabi-Yau shapes there are. Each potential shape implies a distinct pattern of particles and forces. None quite explain the Standard Model.

The sticky-dimension scenario is easier to work with. One idea making the rounds is that the universe is not a single brane, but a framework of multiple branes. Particles are open strings stretched across the framework, forming a giant musical instrument—a multidimensional zither. The arrangement of branes, more than the shape of space, determines the vibrational patterns of the strings. A particle's electric charge and other properties indicate which pairing of branes it connects. Separate branes represent left-handedness and right-handedness, accounting for the mirror-asymmetry of the weak nuclear force.

Particles may be open strings stretching between two branes or two parts of a single brane.

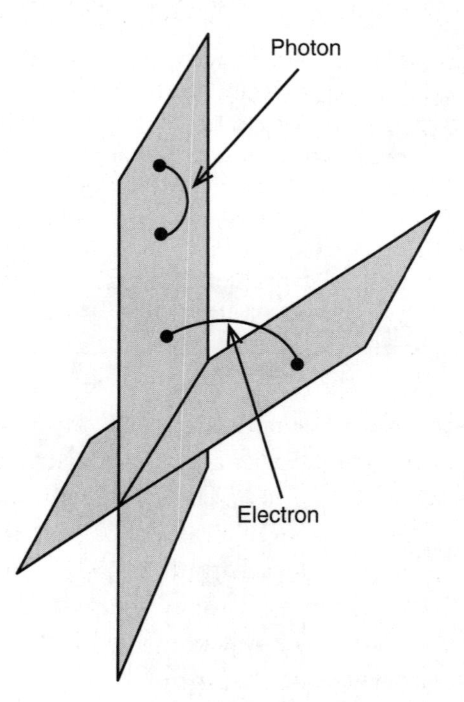

Truth be told, this whole setup of branes is somewhat ad hoc. Eventually, string theorists hope to explain the Standard Model more organically. In the meantime, though, theorists reckon that the best shouldn't be the enemy of the good. Using branes, they make quantitative predictions that experimenters can test at the Large Hadron Collider.

Goldilocks and the Three Dimensions

Why would dimensions come in two varieties, observable and extra? Nobody knows for sure. But one thing physicists do know is that three dimensions are special. Complex life forms seem scarcely possible in two, let alone one or zero. At the other extreme, a higher-dimensional space overly loosens the constraints on objects' behavior, leading to chaos. Planetary orbits through four dimensions go haywire, and atoms can't hold together. The same goes for multiple time dimensions. Apart from giving new meaning to the joke about the man who has two watches and never knows what time it is, multiple time dimensions would make events unpredictable, muddle cause and effect, and destabilize otherwise stable particles.

Quantum Leap

It's fun to speculate whether 2-D life forms could exist after all. In his 1984 book *Planiverse,* computer scientist Alexander Dewdney of the University of Western Ontario came up with some ingenious solutions to the anatomical challenges. Since then, scientists have discovered that 2-D space is surprisingly rich. One of the hottest areas of materials science right now is "plasmonics," in which light behaves as a 2-D wave (a plasmon) rippling along the side of a metal sheet. Experimenters have developed plasmonic mirrors, lenses, microscopes, even lasers. As for gravity, a 2-D space is too constricting to allow bodies to exert gravitational forces on one another. But gravity can operate in other interesting ways, even forming black holes.

As paltry as three dimensions may seem compared to the wonders of higher dimensions, they have the distinct advantage of allowing us to exist. Fewer are too Spartan; more are too permissive; and three are just right. Broadly speaking, there are two ways to explain why we got so lucky to live in a space that's just right for us. The first is that we had no choice. It's the same reason that intelligent life in our solar system arose on Earth but not on the other planets. The universe could be filled with branes of all dimensions in the same way our solar system is filled with planets, and most are either as desolate as Mercury or as menacing as Jupiter.

The second possibility is that the specialness of 3-D space goes beyond the requirements of living things. For instance, simple geometry favors some numbers of dimensions over others. If branes of all dimensions float around out there, they occasionally collide and destroy each other. The smaller the number of dimensions in a brane, the smaller the target it presents. Think of a pot of water on the stove. If you drop in two sheets of seaweed, they're guaranteed to hit. If you drop in two strands of spaghetti, they might. If you drop in two peas, they probably won't. In paring back the number of dimensions, from 2-D (sheets) to 1-D (spaghetti) to nearly 0-D (peas), you reduced the likelihood of colliding foodstuffs. In a 10-D stew, a 3-D brane is more likely to survive than a higher-dimensional one.

Three dimensions also emerge as the preferred number in a couple of the alternatives to string theory. In the "buckyspace" theory, for example, the collective behavior of the building blocks of space give rise to three dimensions over large distances. Why buckyspace works this way, no one quite knows; but it's somehow related to the importance of cause-effect relations. When researchers consider building blocks on which effect can precede cause, their simulations predict that space scrunches up into a tight, chaotic ball.

Quantum Leap

Another tantalizing reason for three dimensions emerges from the foundations of quantum theory. The theory holds that we can obtain only a limited amount of information about an object. For the simplest particle, we can make three mutually exclusive measurements. It seems awfully coincidental that this number matches the number of dimensions of space. Physicist Carl Von Weizsäcker noted this fact in the 1950s, but physicists have yet to explore it much further.

All this discussion skirts several deep questions: Why does the concept of a dimension matter to begin with? Why can the state of a physical object be described in terms of spatial position? And why are a certain number of numbers required to specify this position? These questions go back to the problem of what space and time are. They may not, in fact, be fundamental, but arise from the shenanigans of even more fundamental concepts (see Chapter 18). In that case, the concept of a dimension isn't fundamental, either. When physicists figure out what space really is, the answers about extra dimensions may fall into place.

The Least You Need to Know

- To us, space appears to have three dimensions: length, width, and height.
- The choreography of string theory needs a far grander dance floor: 10 dimensions of space in all, plus one of time.

- We don't see those extra dimensions directly because either they're too small or our usual three dimensions are too sticky.
- Our 3-D space might be a "braneworld" floating through the full 10-D space.
- According to other theories, space does not, in fact, have higher dimensions.

Chapter 15

Parallel Universes

In This Chapter

- What a theory of everything should be like
- The string landscape
- Multiple types of multiple universes
- The anthropic principle

Which of the following would be creepier? An identical copy of you, on an identical copy of Earth, somewhere out in deep space? A nearly identical copy of you, differing only in eye color, but otherwise the same? Or a creature so unlike you, not even having eyes, made up of particles so alien that you could never meet without instant death to you both? These situations sound like science-fiction stories, but your doppelgängers might really exist out there, located in parallel universes that make up a grand multiverse. The idea of parallel worlds is one of the most controversial aspects of string theory. Yet the alternative is weird, too: that our world is preordained right down to the last detail, leaving no scope for historical processes that might turn out differently in different regions of space.

So Many Ways to Make a Universe

It's sometimes said that the ultimate goal of physicists is to revolutionize the fashion industry. They seek to boil the entire physical world down to equations compact enough to silk-screen onto a t-shirt that all the cool kids would wear. Actually, though, physicists' sartorial ambitions are more modest; they hope to create a theory that won't even need a shirt to write it on because it's so blindingly obvious.

Obvious? Given the difficulties they're having in figuring out quantum gravity, it must be one of those things that's obvious only in hindsight. But physicists hope that once they know the trick, all else will follow as a matter of course. They'll find that things are as they are because there's no other way they could be. You'll finally have answers to imponderable questions such as why people talk on their cell phones while driving: it'll be a consequence of the mathematics as surely as the symbol π is.

Over the years, people have put forward philosophical and theological arguments for why the deepest theory should be unique and self-evident. The best evidence, though, may simply be past experience. It's hard to create a consistent theory in any branch of physics because beautiful hypotheses have a nasty habit of collapsing under the weight of their own internal contradictions. In fact, physicists take perverse pleasure from the fact that a quantum theory of gravity is taking so long to figure out. When they do figure it out, they'll have some confidence it must be right because there probably won't be any alternative.

That was the original hope of string theory. It got a boost in the mid-1990s when theorists caught sight of a unique formulation, M-theory. But we have a wee little problem. How does this theory relate to the world we actually see? It involves 10 dimensions of space, yet we see only three. Our world is no more than a pale shadow of the full reality.

Our 3-D hands can cast a huge variety of 2-D shadows on a movie screen. Similarly, as we translate from the glorious unique theory to the messy observed world, the theory becomes progressively less able to predict what we'll see. When we flatten one of the dimensions, leaving nine, M-theory gives rise to not one but five consistent versions of string theory. Continuing, we find hundreds of thousands of ways (or more) to fold up the remaining six extra dimensions. For the extra-dimensional space to hold its shape, it must be pinned down by crisscrossing forces that act like invisible threads. This can happen in so many ways that string theorists have lost count.

In each of these shapes, strings vibrate somewhat differently, leading to a distinct pattern of particles and forces. In one shape, gravity might be so strong that the whole universe is a heaving mass of black holes. In another, observable space might have only two dimensions. In a third, drivers might actually pay attention to the cars around them. String theorists call the full set of possibilities the *landscape*, a term borrowed from evolutionary biology. The name refers to the abstract graph used to represent the full gamut of shapes.

Most of the shapes are unstable, and space thrashes around until it settles into one of the stable shapes, where the natural dynamism of the space is restrained by forces within it. The stable shapes are represented by a "valley" in the landscape. All the valleys seem roughly equivalent. Some are cozier than others, but string theorists have yet to find one that is the clearly preferred resting place for space—so they don't know why our universe is the way it is.

def•i•ni•tion

The string **landscape** is the diversity of possible shapes that the hidden dimensions of space could have. Each shape corresponds to a different set of particles and forces in 3-D space.

Planning for Every Contingency

Skeptics of string theory think this proliferation of shapes is reason enough to dump the theory in the same big pile as all the other beautiful hypotheses that have collapsed over the years. String theorists, however, think the fault lies in their original criterion for success. Maybe they were wrong to expect that the ultimate theory would determine the world uniquely. After all, other theories of physics allow for a multiplicity of outcomes, too.

Consider the classic physics t-shirt that shows the equations of electromagnetism, below which is the caption, "And then there was light." The equations look gnarly but are actually pretty simple once you know how to read them. Based on them, you could describe every light beam that has ever glinted in your eye, every radio broadcast you have ever heard, and every comb that has ever made your hair stand on end. How can one little set of equations do all that? How can simplicity capture complexity?

The answer is that it can't. There's a conceptual division of labor. The equations can determine the behavior of light, but you have to set them up with starting conditions. For a radio broadcast, you must specify the signal, the type of antenna, the atmospheric conditions, and so on. Much of the complexity is tucked inside those circumstances, which have nothing to do with light per se. Scientists and philosophers call them *contingent.*

def•i•ni•tion

A **contingent** situation didn't have to be the way it is. It reflects the way that history unfolded.

In physics as in everyday life, many of the things that seem to happen on purpose really happen by accident. As they say, never ascribe to malice what you can ascribe to incompetence, and never ascribe to design what you can ascribe to luck. People long ago thought the orbits of the planets in our solar system were preordained by geometric principles. It turns out they're largely contingent. Planets around other stars are arrayed differently, reflecting the amount of material in each system, the type of star, the random effects of nearby stars, and countless other circumstances. The laws of physics set general limits but allow for a huge number of detailed arrangements.

Quantum theory adds another level of happenstance. Not only do you need to seed the theory with starting conditions, but you also have to relate its predictions to your observations. An element of randomness rears its head. Physicists think that random quantum processes occurring at critical junctures in the history of the universe set many basic properties of the world. Take the eternal mystery of physics: the fact that toast tends to fall with the buttered side down (see Chapter 2). As physicist Robert Matthews of Aston University in England has shown, the mass of the proton is largely to blame. This particle weighs 1,836.15 times as much as the electron. Why? There's no known reason for it. It appears to have settled into that value at random. Physicists have looked for patterns and ratios among the masses of particles and have found none. It's as though objects, lacking any specific instructions, chose what to do on their own.

In the Loop

Not only is it possible that what we now regard as arbitrary initial conditions may ultimately be deduced from universal laws—it is also conversely possible that principles that we *now* regard as universal laws will eventually turn out to represent historical accidents.

—Steven Weinberg, University of Texas, Austin

In this context, the string landscape fits right in. If planetary orbits and particle masses are contingent, why can't the shape of extra-dimensional space be, too? In this view, the details of the observed particles and forces don't reflect any deep principles of mathematics; they're just the ones we got stuck with. We can draw one of three conclusions:

- **Somebody goofed.** Maybe the landscape is a mirage, and if physicists keep plugging away, they'll find a unique outcome after all. The landscape could have an especially welcoming valley where the observed universe naturally ends up. Or a more sophisticated analysis might find that M-theory or one of the alternatives to string theory explains the Standard Model straight away with no role for contingency.

- **We'll never know why.** Our universe just settled into one shape at random, and that's that. The Standard Model is a *brute fact* or divine imperative beyond our capacity to fathom.
- **We live in one universe among many.** If the ultimate theory doesn't have any way to choose among the possibilities, then they must *all* have happened—every one of them.

def•i•ni•tion

A **brute fact** has no deeper rational explanation. It just is.

Time will tell which of these is right. For now, the third option looks the most promising scientifically. It's very hard to write down a theory that generates everything we see and nothing besides. Our observable universe may be just one of countless universes within a space so vast that your brain will explode all over this book and create a hideous mess if you try to imagine it. Parallel universes have jumped out of science-fiction novels into textbooks, and the real version is even more mind-bending than the fictional.

Making the Possible Real

People have long speculated about multiple universes, but modern cosmology makes them more plausible than ever. Various processes naturally create them. We don't even need string theory. Current theories are quite enough. According to relativity theory, for example, space tends to grow and pinch off to form new regions. It'd be a bigger mystery if other universes *didn't* exist. Something would have to restrain space from doing what comes naturally.

Cosmologists used to define "the universe" as "everything there is." But nowadays they distinguish between *our universe* (the part we can directly see), other universes (volumes of space comparable to ours), and the *multiverse* (an entire collection of universes). We even have different types of multiverses. Cosmologist Max Tegmark

def•i•ni•tion

The observable universe, or simply **our universe,** is the sum of all we can see, which is not necessarily all that's out there.

The **multiverse** is the larger grouping of which our observed universe is only a part.

of the Massachusetts Institute of Technology classifies them in four levels, which get progressively more expansive.

Level 1: Space Beyond Our Horizon

From the top of the Empire State Building, you can see about 50 miles. As much as New Yorkers may like to think they're the center of the world, even they wouldn't claim that Earth stops at their horizon. Similarly, astronomers can see about 47 billion light-years, which is as far as light has traveled since the start of the big bang (taking into account the expansion of space). But who's to say space stops there?

How far does it go? From the Empire State Building, you could in principle survey the vista to deduce Earth's curvature and therefore its size. Astronomers, likewise, have measured the curvature of the universe by surveying the cosmos. Specifically, they look for signs that light rays traversing the universe bend in a long arc. Such bending would have a funhouse effect, causing distant galaxies and pregalactic scraps of matter to appear abnormally large or small. Nothing like that has shown up. If the universe is a giant sphere, it must be at least a thousand times bigger in volume than the current universe.

All Tangled Up

Could the universe be smaller than it looks? If it had a curious doughnut shape, light wrapping around it could produce the illusion of immensity like a mirrored elevator or dressing room. But then astronomers should see repeated patterns of celestial bodies, and they don't. By the best current measurements, even a doughnut-shaped universe must be at least 40 billion light-years in radius.

So at least a thousand parallel universes are out there beyond the range of our vision. The number could well be infinite. Those other universes should be pretty much the same as ours. The main difference is their initial endowment of matter, which was set by random processes at the dawn of cosmic history. Over billions of years, the varying matter endowment led to diverse arrangements of galaxies, stars, and planets.

If space is infinite, things get truly mind-blowing. There are an infinite number of planets, an infinite number of Earths, and an infinite number of people who go by your name. So what distinguishes fact from fiction? Every story you can tell, as long as it doesn't violate the laws of physics, occurred for real on one of those other Earths. Everything that can be, is.

Level 2: Bubble Universes

On the vastest scales, not only does the initial endowment of matter vary, but even the observed laws of physics could as well. The key is the process of cosmic inflation, the explosive expansion that our universe seems to have undergone long before stars and galaxies formed (see Chapter 9). Cosmologists think that inflation is actually the default mode of space. The high-energy conditions at the dawn of time ignited it, and it perpetuated itself in a chain reaction. In some isolated bubbles, though, inflation burnt itself out, and matter had a chance to gain a foothold. Our universe and its neighbors were one result. There's nothing to stop other bubbles from emerging, too. As long as the space between bubbles grows faster than new ones nucleate, inflation goes on forever and spawns an infinite number of bubbles.

Quantum Leap

Theorists speculate that bubble universes could pop up all over the place. The extreme conditions in a black hole, the irrepressible fluctuations of quantum fields, the high pressures in the laboratory of an advanced civilization, and the implosion of a universe might all reignite inflation and give birth to a baby universe.

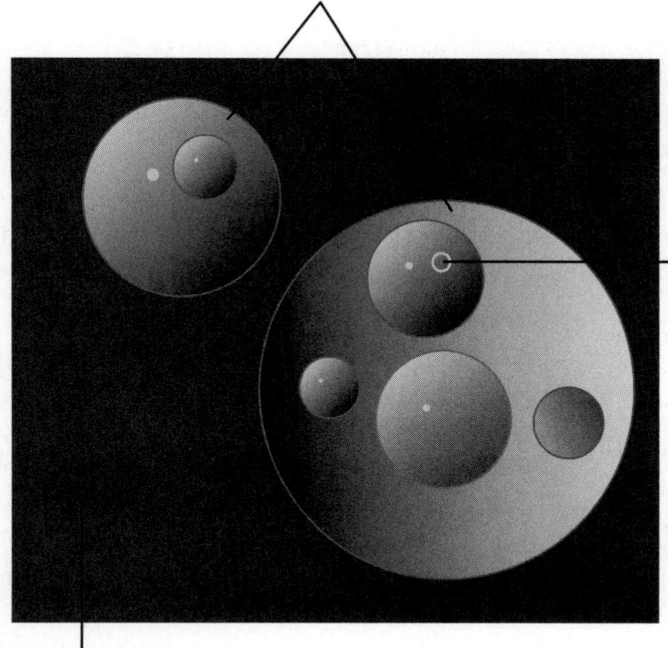

Our observable universe may be just one patch of a much larger universe or multiverse.

According to string theory, when a bubble first forms, the space within it has an unstable shape. The energy released by space as it thrashes around pumps up the volume of several of the dimensions (see Chapter 20). Inflation ends as space settles down. (It can resume later because the default mode of space, even in the stable shape, is to expand.) But each bubble settles down in its own way, leading to different sets of particles and forces. Even the number of visible dimensions could vary. In this way, inflation turns the full landscape of possibilities into reality.

In our bubble, the proton is 1,836.15 times heavier than the electron. In another, it might be 459.038, which, all else being equal, would cause toast to fall with the buttered side up. Alas, you'll never be able to venture into that happy world of clean kitchen floors. Even if you managed to escape from our bubble, the workings of particles and forces would change as you entered the lifeless void outside it. No spacesuit could protect you from a metamorphosis in the very laws of physics. It wouldn't be pretty.

Level 3: Quantum Many Worlds

In the movie *Sliding Doors*, Gwyneth Paltrow's character either catches a subway train or misses it. The film follows the two alternative ways her life could have unfolded, a happy one with a tragic ending and a sad one with a more hopeful ending. The film captures an important aspect of quantum theory: everything not forbidden is compulsory. All the possible outcomes predicted by the theory behave as if they actually happen. The theory treats them on an equal footing; it doesn't single one out. The possibilities even affect one another through the behavior of quantum waves, which they wouldn't if they were purely hypothetical.

That's as true in string theory as in standard quantum theory. All the possible shapes of extra-dimensional space behave as if they actually happen. If something behaves as if it's real, the most straightforward conclusion is that it *is* real. So we don't even need cosmic inflation to turn the string landscape into reality.

We can think of the quantum possibilities as parallel universes existing not in ordinary space but in a more abstract realm. Apart from their location, they are identical to the level two parallel universes. In fact, some physicists put two and two together. They argue that quantum theory is really a theory of parallel universes. This point of view, known as the *many-worlds interpretation*, holds that there's no distinction between what's possible and what's real. Possibility is just existence in one of the parallel universes.

def•i•ni•tion

The **many-worlds interpretation** of quantum theory is a way of making sense of quantum randomness in terms of the existence of parallel universes.

Quantum Leap

Suppose you had a quantum coin, one governed by quantum effects rather than ordinary effects in the way you flip it. (A particle can act like such a coin.) Quantum theory says that after you flip it and catch it, but before you peek, it's both heads and tails at once. That sounds mysterious but makes perfect sense if the coin is heads in half the parallel universes and tails in the other half. If you see heads, that just means you're sitting in a universe in the first group. So the many-worlds interpretation resolves one of the nagging paradoxes of quantum theory: why we see a specific outcome even though the theory doesn't single one out.

Level 4: The Mathematical Universe

There's an even broader conceivable type of parallel universe, not just with different allotments of matter or particle types, but with different fundamental laws. If M-theory rules our universe, Newton's laws might run the show in another, and Pythagorean geometry in a third. This idea comes in multiple versions. Tegmark has argued that the laws must at least be mathematically consistent. Others, such as the late American philosophers Arthur Lovejoy, Robert Nozick, and David Lewis, threw caution to the wind and speculated that absolutely anything goes.

It sounds awfully weird. But what other options are there? If there aren't level four parallel universes, then what led nature to choose the laws that govern our universe? Are they the only laws that are internally consistent, or are they a brute fact? Whichever it is, I think we can safely bet that the answer will be weird.

Types of Multiverses

Level	Parallel Universes	What Varies	Hollywood Pitch
1	Regions beyond our range of vision	Initial density of matter	*Groundhog Day* meets "The Library of Babel"
2	Bubbles in otherwise empty space	Types and properties of particles and forces	*A Tale of Two Trillion Cities* meets *Worlds Without End*

continues

Types of Multiverses (continued)

Level	Parallel Universes	What Varies	Hollywood Pitch
3	Quantum possibilities	Types and properties of particles and forces	*Sliding Doors* meets *Permutation City*
4	Everything that is possible at all	Even the fundamental equations	*The Matrix* meets *Finnegan's Wake*

The Best of All Possible Worlds

What appeals to physicists about parallel universes is not just that string theory and other theories predict them or that they make for great TV and movie plotlines, but that they explain certain otherwise baffling aspects of our universe. The Standard Model is filled with numbers such as particle masses that look as if they were chosen at random. Physicists can discern only one common thread running through them: their values are connected to our own existence.

If these quantities had been much different from what they are, our universe might have become a thin gruel of subatomic particles or a dense soup of black holes. A shift of just a few percent in the strength of electromagnetism or the strong nuclear force would have created stillborn stars and scarcities of must-have elements such as hydrogen and carbon. Not only do the quantities have the right values to allow for our existence, but they also ensure that intelligence has some survival value. Nature may sometimes be capricious, but for the most part, it abides by rules we can grasp, and it didn't have to be that way.

Entire books and websites have been written on these *anthropic coincidences.* Wikipedia has a good list of them. The multiverse neatly explains them. There's a huge range of universes out there, and we simply happen to live in a hospitable one. The idea sounds more palatable if we replace the word "universes" with "planets." If Earth were the only planet in the entire universe, its habitability would be either a brute fact or a necessary consequence of the laws of physics. Neither is true. Astronomers know of hundreds of planets, and our galaxy alone could have billions, coming in a wide range of orbits and sizes. We live on a cozy one just because there's no place else we could.

This kind of argument is known as the *anthropic principle.* At the very least, it cautions scientists that what they see may not be a representative sample of what is. Just because our planet or universe is comfy doesn't mean it had to be that way.

def•i•ni•tion

Anthropic coincidences are the fortuitous values of many crucial quantities in nature. Without these values, humans couldn't exist.

The **anthropic principle** is the concept that we have to take our own existence into account when trying to explain the universe.

The anthropic coincidence that has gotten the most traction concerns the notorious dark energy, the unknown form of energy that is causing the expansion of the universe to accelerate (see Chapter 9). Dark energy laid low for the vast bulk of cosmic history but then began to exert its influence fairly late in cosmic history—around the time, in fact, that our solar system formed. If it had arisen much earlier, it would have scattered material before galaxies had a chance to agglomerate and stars had a chance to rev up, and we wouldn't be here. Consequently, the density of dark energy must be incredibly small, just a hair above zero. Yet its most natural value is either exactly zero or extremely large (see Chapter 7).

According to string theory, dark energy arises from the same process that inflation did—namely, the shape of extra-dimensional space. The level two parallel universes have varying amounts of dark energy. The huge number of those universes implies a huge number of permissible amounts of dark energy, and that's just what we want. The more universes there are, the more likely one of them will have the right amount of dark energy to explain the coincidence.

That said, not all properties of our universe make sense in terms of the multiverse. An example is the asymmetry of matter and antimatter. Matter didn't have to be so overwhelmingly dominant for us to exist. Cosmologists estimate that the relative amounts of matter and antimatter could vary by a factor of 10 million without making the universe inhospitable to life. There's no uncanny coincidence to explain. Maybe this is a property that happened on purpose rather than by accident.

Another example is the orderliness of the universe. Compared to what it could be, the cosmos is a neat freak. This is what differentiates past from future. Order naturally degenerates into disorder, whereas disorder just stays that way. For time to march on, it can't have reached the fully disordered state yet. Yet the universe is far more orderly than it needs to be for us to exist.

The anthropic principle remains very controversial. Its detractors think that all the quantities of physics will ultimately prove, like the dominance of matter or the orderliness of space, not to be so coincidental after all. To them, appealing to a multiverse is little better than chalking it all up to a brute fact. Both answer "Why?" with a bleak "Because." But detractors and proponents agree on one thing. Physicists need to keep looking for a unique explanation of our world because if they don't look, they'll surely never find it.

Some say that an endless universe makes human life seem so insignificant. I think the opposite is true. It shows what a special place our Earth really is. If our planet had been an inevitable consequence of the laws of physics, it would cheapen the accomplishment that it is. Our world stands out all the more for the vast lifeless void that surrounds it.

The Least You Need to Know

- Arguably the central question in physics today is which things are hard-wired into the laws of nature and which are accidents of circumstance.
- String theorists originally expected everything to be hard-wired but now think that almost everything is accidental.
- Things may have played out very differently in other regions of space, creating a multiverse of parallel universes.
- The only explanation for many features of nature may be that they are preconditions for our own existence.

Chapter

The Root of the Tree

In This Chapter

- Spacetime foam and its effects
- What strings do as they shrink
- The possibility of a fractal space
- Making relativity extra special

All candidate quantum theories of gravity say the tree of physics has a root, a limit to how small things can possibly be. And whenever someone sets a limit, the first thing people want to do is test it. So what happens if you try to make something even smaller? Space and time become elusive concepts and the very notion of "smaller" loses meaning.

Swallowing Its Tail

In Chapter 6, I mentioned the 1957 film *The Incredible Shrinking Man* and commented that, if the studio had made a sequel, the hapless protagonist would have confronted adventures far more mind-blowing than dueling with a spider. He might have plucked strings (and I don't mean guitar strings), rambled through extra dimensions, and watched quantum particles make up their minds about which way to go. But not even the most inventive screenwriter and FX team could have kept the franchise going past *Episode III: Revenge of Gravity.* According to all proposed quantum theories of gravity, our hero could only shrink to a certain minimum size.

At first, as he shrinks from his normal size, gravity fades away. The little man floats in the air like a dust mote, and forces such as electromagnetism dominate his struggle for survival. But as he reaches sub-subatomic proportions, gravity reappears. For instance, consider what it takes just to see him. You need to use photons of ever-shorter wavelength to illuminate him. Shorter wavelengths mean higher energy and, if the energy is high enough, photons do more than exert an electromagnetic force. They also exert a noticeable gravitational force and contort the very things they seek to reveal, offsetting the higher resolution provided by the shorter wavelength. The image of our hero starts to look like one in a funhouse mirror, and shorter wavelengths of light only worsen the distortion.

At a certain point, the gravitational effect becomes dominant. Continuing to increase the energy then becomes counterproductive because it only reduces the resolution of the image. The minimum possible distance you can resolve is the Planck scale of 10^{-35} meter (or a bit greater, once you account for the gravity exerted by the measurement apparatus, too). If the shrinking man manages to reach this size intact, he becomes the invisible man as well. He can't even make out his own hand.

That's the least of his worries. At fine scales, the quantum fields all around him fluctuate so tumultuously that their gravity becomes a force to be reckoned with. Tiny black holes form and pop like soap bubbles. The fluctuations might rip the spacetime fabric and create wormholes, in which case spacetime ceases to be a static continuum and becomes a bubbling froth of interconnections, called *spacetime foam.* The concept of being at a specific location loses meaning as our hero gets lost in black holes or slips into wormholes and rides through them like a roller-coaster.

def•i•ni•tion

Spacetime foam is the hypothesized frothy behavior of spacetime on the smallest scales, caused by quantum fluctuations of the gravitational field.

In these ways, the Planck length acts like the smallest possible tick mark on a ruler. You can't locate an object with any greater precision that that. Gravity sculpts the universe not just on its largest scales but also on the very smallest ones. To depict this dual role, physicists Sheldon Glashow and Joel Primack have adopted the metaphor of the uroboros, a classical mythological symbol of a snake eating its own tail.

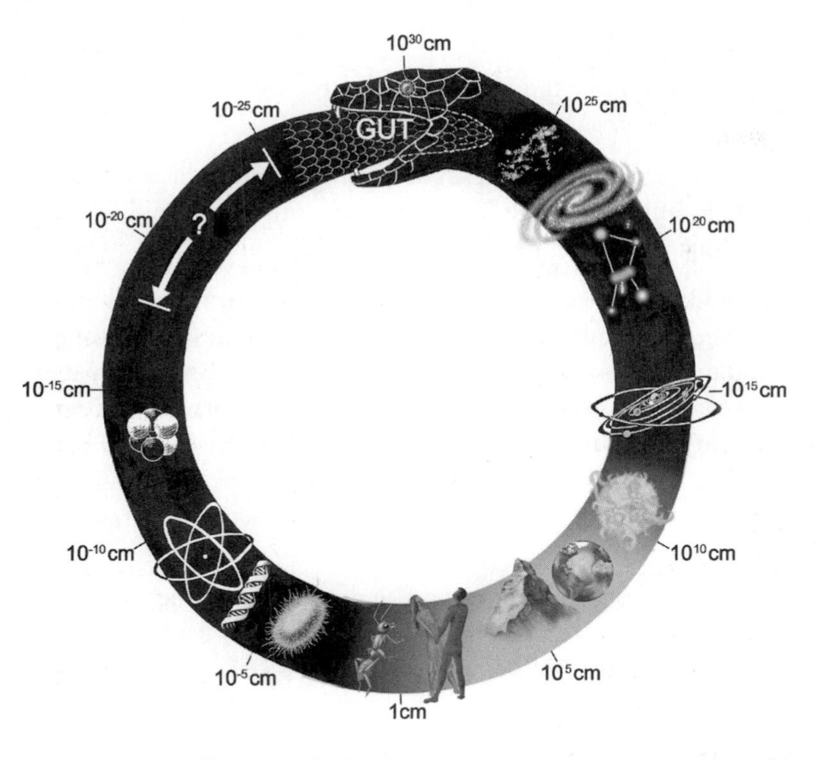

The cosmic uroboros shows the connections between very large and very small.

(Copyright 2006 Joel R. Primack and Nancy Ellen Abrams, from The View from the Center of the Universe*)*

Quantum Leap

The cosmic uroboros shows the vast range of size scales in nature, spanning over 60 orders of magnitude from the smallest conceivable distance (the Planck scale) to the largest observable distance (the radius of our universe). We humans are nearly in the middle. To be precise, the center is closer to a paramecium, but calling a paramecium the center of the universe sounds like a Gary Larson cartoon. That's no coincidence. We're big enough that quantum effects don't make our lives unpredictably random and small enough that gravity doesn't crush all our intricacy. The observable cosmos has to be as big as it is so it could have cooled enough for complex structures to form.

String theory adds some new twists as it reveals other connections between the small and the large. The photons you use to illuminate the shrinking man are actually little strings, and as their energy increases, their intrinsic size (typically estimated to be 10^{-34} meter) becomes a factor even before gravity does. The strings distend and become less able to pick up fine features. Eventually increasing their energy starts to reduce the resolution of your image. You reach a boomerang point where you think you're probing shorter distances but actually are probing longer distances again. The snake doesn't so much eat its own tail as fold back on itself. If you prefer an arboreal theme, the tree of physics flips upside-down like the Hindu banyan tree that has its roots in the heavens and its branches in the ground.

The ability of short distances to act like long ones is surprising because the infinitesimal and the infinite don't normally have much to do with each other. Yet unification is all about making connections where you never thought there would be any.

Alternatives to string theory likewise predict that our notions of spacetime break up on small scales. The ways theories handle the roots of the tree of physics is one thing that distinguishes them and might make them experimentally testable. Any time particles or strings are pushed together cheek by jowl, as they are within black holes and early in cosmic history, the fine structure of spacetime can have big consequences.

Living Off the Grid

A minimum distance suggests that spacetime has some kind of cellular structure. Might it be a regularly spaced grid like a chessboard? In that case, you could no sooner exist in the blank space between grid points than a chess piece could sit halfway between two squares. The smallest meaningful distance is the distance between adjacent points.

Unfortunately, such a grid is hard to square with Einstein's special theory of relativity. According to this theory, people moving at different speeds perceive distances differently, so only some of them would see the grid as regularly spaced. Their viewpoint would be singled out as uniquely valid, as if the universe had been built specially for them. Apart from the unfairness of it all, privileging one viewpoint could allow some people to see an effect precede its cause.

That might not be such a bad thing if the violation became apparent only on fine scales, but any violation should be amplified by distance. It would take a new principle, yet to be discovered, to isolate a failure of relativity on small scales from its undeniable success on large ones. Some physicists have sought such a principle, as I discuss later, but most think special relativity holds all the way down. That demands a more subtle view of spacetime than a grid of points.

Dualing Points of View

String theory is nothing if not subtle. In it, the shortest distance is not a property of spacetime per se but of the strings that propagate within it. The processes that govern strings are defined not in ordinary spacetime but in the inner space of the string (see Chapter 12). When you ask what the string does in the external spacetime, some useful ambiguity arises. Two entirely separate modes of behavior can be outwardly equivalent or dual. *Dualities* are the physics version of optical illusions that consist of one image with two interpretations: a vase or two faces; a cube oriented down and to the right or up and to the left; a saxophone player in profile or a woman's face. Both ways of viewing these images are equally valid.

def•i•ni•tion

A **duality** is a relationship between two situations that look utterly different yet are actually equivalent.

Dualities crop up all over physics. For example, electric fields and magnetic fields are dual: in many situations, you can ascribe an electrical effect to a magnetic field, and vice versa, without changing a thing. String theory is a veritable jungle of dualities. Many of them cause short distances to masquerade as long ones and vice versa. In one case, a closed string trudging a long distance looks like a pair of open strings navigating a short distance. When viewed in spacetime, both situations sweep out the same tube. The only difference between them is what you call "space" and what you call "time." If time runs lengthwise, you have a closed string moving a long distance. If time runs around the circumference, you have a balanced pair of open strings coming into existence, moving a short distance, and then canceling each other out. According to special relativity, these two interpretations are equivalent.

Another duality, known as *T-duality*, involves the size of the space that strings move in. For the sake of argument, imagine that space has a cylindrical shape and that the string glides around the circumference. If the circumference is large, the string barely notices it's on a cylinder at all. By giving it a push, you can send it off at pretty much any velocity you like.

def•i•ni•tion

T-duality is the equivalence between a small extra dimension and a large one.

If the cylinder is narrow, the limited room for maneuver affects the string's motion. Quantum theory describes its motion in terms of a wave, the wavelength of which encodes the string's momentum. By restricting the wavelength, the cylinder size restricts the velocity. The string becomes like a car with a malfunctioning cruise control that only lets you set the speed in increments of, say, 8 mph. You can make it go 64 mph or 72 mph but not 65 mph. The smaller the circumference, the bigger the increment becomes.

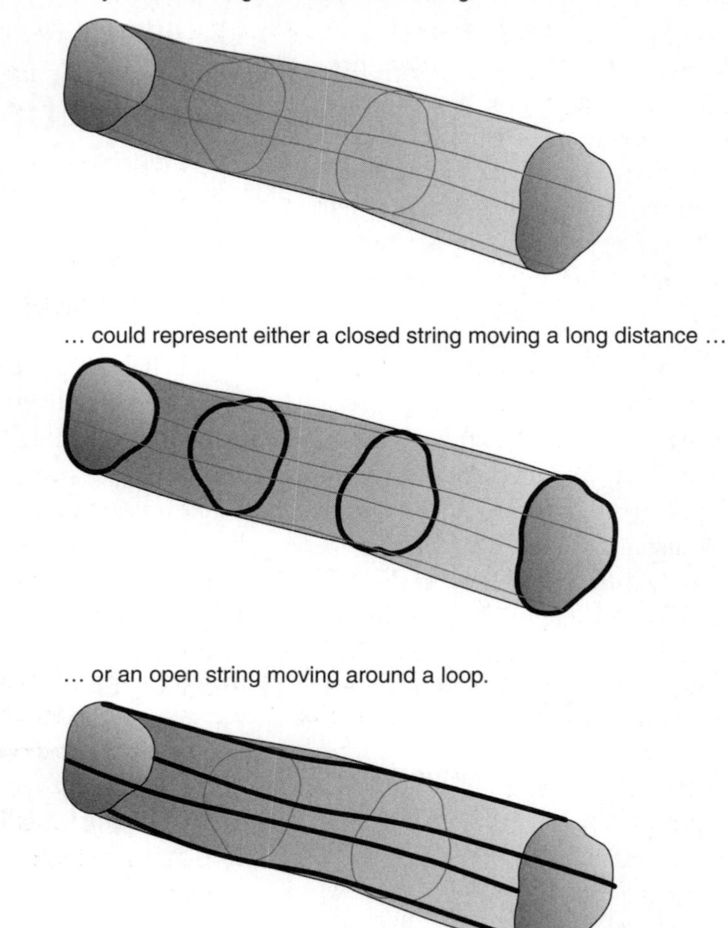

Just as its motion is becoming constricted, the string gains the freedom to behave in a peculiarly stringy way: wrapping itself around the cylinder like a rubber band around a rolled-up poster. A rubber band wrapped around a poster multiple times stores a lot of elastic energy. The same goes for a string wrapped around a cylindrical space multiple times. (One difference is that you don't have to pull the string off the end of the cylinder to add a winding. It breaks and reconnects to form the loop.)

For a large circumference, adding a winding represents a huge step up in energy; for a small circumference, it's just a small step. That's exactly the opposite trend from motion. For large cylinders, it takes less energy to accelerate than to wrap, but for small ones, it takes less energy to wrap than to accelerate. Those of us watching the string can tell only what its overall energy is. We don't know whether the energy represents motion, winding, or vibration (which is insensitive to the size of space). We can't distinguish between a closed string moving around a large cylinder and a closed string wrapping around a small cylinder.

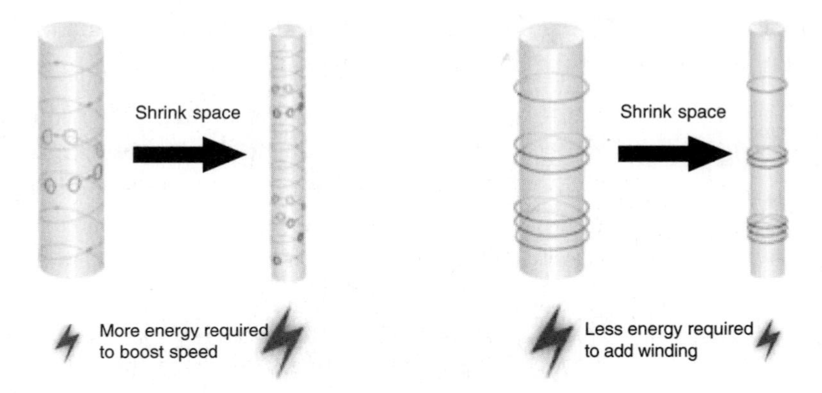

As you shrink space, it becomes harder to increase the speed of a moving string but easier to wrap it around space.

This ambiguity arises not just for closed strings moving around a cylinder but also for closed and open strings in other settings. The crossover point, where motion and wrapping are equivalent, is perhaps 10^{-34} meter. To an outside observer, this represents the minimum size of space and therefore of the string that resides in it. The transition from motion to wrapping occurs gradually, which avoids the contradictions of a grid-like space. Instead of a chessboard, space looks more like a watercolor painting where the brushstrokes bleed together.

Is Anything Smaller Than Strings?

But what about the gap between the minimum string size and the Planck scale of 10^{-35} meter? Those shorter distances fall between the cracks of strings, yet theorists think they still exist. To probe them, you can turn to 0-branes, the exotic particles mentioned in Chapter 12. These 0-branes anchor the ends of strings but can do things on their own and might even be the driving force for string behavior. They are so heavy they can probe short distances without getting distended.

Quantum Leap

Some researchers argue that strings ultimately consist of space-time foam. They may be like rivers that look like smooth currents from afar but are burbling brooks up close.

On these scales, the standard way of describing positions—using one coordinate for each dimension, such as latitude and longitude in 2-D—is hopelessly crude. Instead, it takes a whole array of numbers. These arrays specify the distance between every 0-brane and every other 0-brane, like one of those mileage charts in a road atlas. In principle, you could sit down with a mileage chart and work out where the cities must be located to account for the distances between them. But in the microscopic world, an array of distances may not be so easily reduced to points with definite locations.

If you insist on using standard coordinates, some information gets lost. The more precisely you know the longitude, the less precisely you know the latitude, and vice versa. The coordinates aren't fully independent of one another. This tradeoff is known as *noncommutative geometry.* Some researchers have even sought to use it as the basis of a fundamental theory independent of string theory.

> **def•i•ni•tion**
>
> In **noncommutative geometry,** the coordinates of points aren't fully independent of one another.

Combined with the tradeoffs of standard quantum theory, noncommutative geometry predicts that a fast-moving string oozes out in the direction perpendicular to its motion. For example, suppose your cat is sprinting down the hall at a high but uncertain speed. According to quantum theory, you may not know her speed, but at least you know where she is. Bringing in noncommutative geometry, the well-defined position in the direction of her motion corresponds to a spread of her lateral position. A high-energy cat seems to fluff out.

Loops, Trees, and Sprinkles

The leading alternatives to string theory predict different ways that nature can have a smallest scale while respecting special relativity.

- **Loop gravity.** According to this theory, lengths, areas, and volumes come in discrete chunks the size of the Planck scale—the atoms of space. If you try to probe shorter distances by creating higher-energy photons, space itself resists you. Gravity, normally an attractive force, becomes repulsive and pushes out the excess energy. Although these atoms are interconnected like a grid, it is no ordinary grid. By the dictum that "everything not forbidden is compulsory," space is in all its possible configurations at once. We don't directly see this geometrical goulash; all we see is the average of the myriad possibilities. People moving at different speeds take different averages but still perceive the space atoms to have the same size. So the laws of physics work the same for all, in keeping with the principle of relativity.

- **Buckyspace.** Although proponents of this technique envision spacetime as a framework like a geodesic dome, that's just an approximation that lets them simulate it on a computer. As with loop gravity, the observed shape is a smooth quantum average. The computer simulations have reached two interesting conclusions. First, you need to wire in special relativity at the microscopic level, as in string theory. Otherwise the goulash of possible geometries never boils down to the four dimensions we know and love. Second, spacetime totally changes in character on short distances. It shades from a 4-D continuum into a 2-D *fractal.* Coastlines, trees, and Jackson Pollack paintings are examples of fractals. Each little piece of one looks much like the whole thing; the distinction between scales fades away. Things smaller than the Planck scale merely reproduce, in miniature, the Planck scale. Space doesn't boomerang on itself so much as stutter.

def•i•ni•tion

A **fractal** is a geometric figure built from a shape that repeats on multiple scales.

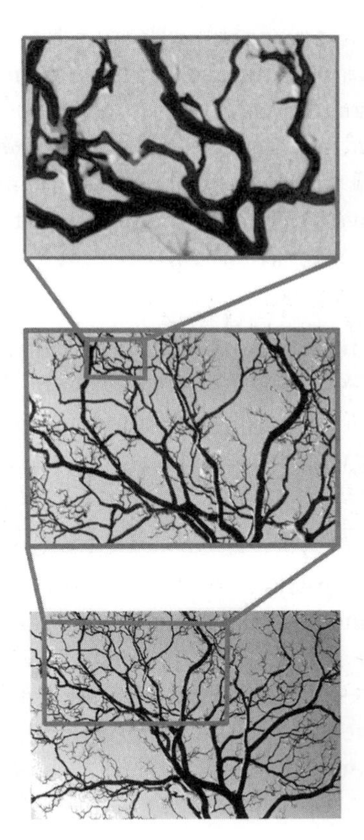

On forest scales, the image of a tree is 2-D; on twig scales, 1-D; and in between, it's a fractal—neither 2-D nor 1-D, but something with qualities of both.

(Courtesy of Richard Taylor)

- **Causal sets.** This technique starts from the premise that spacetime is a grid, just not a regular one. Points are sprinkled at random distances from one another, which ensures that no observer's viewpoint is truer than any other's. Their average spacing is the Planck length. The points are so dense that variations in their spacing are all but undetectable. They interconnect to ensure that cause always precedes effect, in accordance with relativity.

Does Relativity Fail?

However weird the preceding ideas may be, you can take comfort from the perseverance of special relativity. It vindicates some of your basic intuitions about, for example, cause-effect relations. If you see A cause B, I do, too, no matter how fast I'm moving or how small I am. That way, we can come to some reasonable agreement about how the world works. Some physicists, though, don't think even that's the case. They suspect that special relativity either breaks down or takes on a new form in the microworld.

Giving up special relativity would so drastically change the character of string theory that most physicists wouldn't even call it string theory anymore. It would take away most or all of the extra dimensions of space, which were hypothesized to keep special relativity happy. Good riddance, say proponents of this approach, known for technical reasons as "Liouville" strings. They see those dimensions as an unnecessary complication. In return, however, they must assume spacetime foam behaves in an irreducibly random way. For better or worse, this anarchy violates the principle of determinism—everything that happens, happens for a reason.

Another idea merely tinkers with special relativity. A central feature of relativity is the universal speed limit—namely, the speed of light. It's universal because everybody sees light moving at the same speed no matter what his own rate of motion is. It's a speed limit because if you try to accelerate a spaceship from sublight to light speed, you fail. As you approach light speed, your efforts to accelerate achieve diminishing returns. Reaching light speed would require infinite energy.

Maybe the Planck scale plays a similar role. It could be a universal length or energy limit. Although relativity normally says that people moving at different speeds perceive distances differently, maybe this doesn't hold true for a Planck-length object. If one person measures an object to be the Planck length in size, then everyone will. If you try to accelerate an elementary particle to the Planck energy, you fail. As you approach the limit, your efforts to accelerate achieve diminishing returns.

Some physicists have tweaked relativity to include this second limit, creating *doubly special relativity* or *deformed special relativity.* In two-dimensional space, loop quantum gravity predicts just such an effect. Theorists debate whether it happens in 3-D, though.

These proposals are controversial but have the virtue of making distinct observational predictions. Special relativity would begin to make erroneous predictions at very high energies as either the randomness or the universal energy limit made itself felt. Most prominently, the speed of light wouldn't be the constant that Einstein took it to be. High-energy photons might travel more slowly or quickly, depending on the specific theory. Having such a short wavelength, these photons are more sensitive to the spacetime foam, just as a small speedboat is buffeted by ocean waves more strongly than a supertanker is.

def•i•ni•tion

Doubly or **deformed special relativity** is a modified form of Einstein's special theory of relativity that adds a second limit in addition to the speed of light.

For our shrinking man, this would have some dramatic consequences. Space would become a prism, splaying light into a rainbow. The smallifying machine would have to apply increasing amounts of energy to shrink him, and his ordeal might finally come to an end.

The Least You Need to Know

- There's a limit to how small things can be, as if spacetime had some kind of cellular structure.
- String theory suggests that microscopic space is like a watercolor painting where brushstrokes bleed into one another.
- Loop gravity suggests it's a quantum stew of possibilities, and the buckyspace theory suggests it's a fractal.
- In some approaches, the speed of light might change at very high energies and small scales.

Chapter 17

Symmetry

In This Chapter

- What physicists mean by beauty
- The concept of symmetry
- What string theory says about symmetry
- Is symmetry all it's cracked up to be?

As motley as the universe is, order and regularity—in a word, symmetry—underlie it all. The concept of symmetry is central to current theories of physics. String theory both giveth and taketh. It's the most symmetrical theory ever devised, yet it suggests that symmetries physicists now regard as fundamental actually aren't.

Beauty Is Deep

When physicists wax eloquent about a theory, their favorite adjective is "beautiful." So the theory doesn't just explain the world; it does so with style. The pursuit of beauty is as essential to physics as it is to art, poetry, and speed dating. What physicists mean by beauty closely matches what poet Ralph Waldo Emerson wrote: "We ascribe beauty to that which is simple; which has no superfluous parts; which exactly answers its end;

which stands related to all things; which is the mean of many extremes." A beautiful theory explains so much with so little; it clicks together without having to be forced.

Beauty obviously means different things to different people, and discussions of it can start to sound like nauseatingly sentimental '70s love songs. From Dadaism to punk, entire artistic movements have adopted the creed that prettiness comes at the expense of authenticity. In science, skeptics worry that "beauty" just means "comfort level." But no one doing physics can be totally immune to the charms of beautiful things. In the early stages of developing a theory, before physicists have tested it experimentally—and that applies to all the candidate quantum theories of gravity—mathematical aesthetics is all they have to go on. Beauty may be an imperfect guide, but its success rate softens even the hardest heart.

The quality that physicists most often associate with beauty is *symmetry*. It's easiest to think of symmetry in terms of geometric figures, but the concept extends to more abstract concepts. A symmetrical shape remains the same even if transformed: reflected, rotated, or rescaled. A symmetrical equation remains the same even if transformed, for instance, by altering the mathematical variables. An example is the equation for a simple parabola, $y = x^2$. If you replace x by $-x$, the equation doesn't change, which is an abstract way of saying that the left side of a parabola is the mirror image of the right.

def•i•ni•tion

Symmetry is a property of an object or equation whereby it doesn't change even when transformed in some way.

Symmetry has several related qualities:

- **Simple.** It's easier to describe a parabola than a random amorphous blob.
- **Unified.** Symmetry melds parts into a whole. It relates your left hand to your right, the sides of a hexagon to one another, and the notes in a musical chord to one another.
- **Self-explanatory.** A symmetrical body has an inevitability about it. Among the full continuum of possible shapes, only a privileged few are symmetrical. None of their parts could be dropped or changed without breaking the symmetry.
- **Objective.** A symmetrical object stays the same even if we view it from different angles. It's independent of us.

Symmetry plays the same role in physics as a national constitution plays in politics. It is not a law in itself but a set of standards for laws, requiring that laws be the same for everyone despite the vagaries of circumstance. Equations must have built-in mechanisms to correct for our individual points of view and our arbitrary conventions. Often symmetry restricts the equations so much that it determines them almost completely, suggesting that there are only a few ways (maybe just one way) for the world to be put together consistently.

Albert Einstein elevated symmetry to this role because his special theory of relativity embodies a symmetry of motion: the laws of physics shouldn't depend on how fast a person is going. For the laws to satisfy this requirement, they must treat space and time in a unified way. The distinction between space and time and even between electric fields and magnetic fields is like the distinction between left and right—a question of perspective (specifically, your velocity).

As powerful as symmetry is, we don't want too much of a good thing. In physics as in art, the greatest beauty combines harmony with dissonance, structure with surprise. A perfectly symmetrical world would be a sterile one. In the parable of Buridan's Ass, a donkey has two bales of hay—one on the left, one on the right. The poor thing can't decide which to eat. Either it breaks the symmetry, or it does nothing and starves. A similar dilemma plays itself out every day on city sidewalks as pedestrians walk toward each other. They have to break left or right. Two people, locked in the perfect symmetry of listening to their iPods, are going to make asses of themselves.

So it goes for nature. The symmetry of the laws of physics reflects the equivalence of various possible situations, but we experience only one of those situations. What makes the choice is one of the central questions in modern physics.

Types of Symmetries

Symmetry isn't just a pleasing egalitarian ideology. It has tangible consequences. One (see Chapter 2) is the fact we can't create or destroy energy. The conservation of energy has to do with a symmetry of time or, if you like, the symmetry of procrastination. If we put off something to tomorrow, all else being equal, the outcome will be the same. If two particles smash together in a certain way today, they'll smash the same way tomorrow. Their combined energy couldn't change in the meantime without spoiling this equivalence. Sadly, the all else being equal part is easier to achieve for particles than for humans.

Another symmetry lets birds perch safely on bare high-voltage power lines. Only if birds straddled the wires and the ground, creating a difference of thousands of volts across their little bodies, would they fry. The absolute value of the voltage doesn't matter; only voltage differences do. In the equations of electromagnetism, we could add 100,000 volts to every voltage and nothing would change.

Because of this, the energy required to create a particle can't depend on the absolute voltage. That has an interesting implication. Suppose we could create an electron at 100,000 volts, let it drop to zero volts, and then destroy it. The creation and destruction would be a wash, but the fall in voltage would release energy. The net effect would be to create energy, and that's never seen. So we can't, in fact, create an electron on its own. We always need to counterbalance it with a positively charged particle, keeping the net charge fixed. Just as the symmetry of time implies the conservation of energy, the symmetry of voltages implies the conservation of electric charge.

These two cases are the archetypes of the two types of symmetries:

- **Spacetime symmetry.** Concerns external properties of particles such as their position, velocity, and orientation.
- **Internal symmetry.** Concerns internal properties of particles such as their electric charge or quark color.

Each of these two types, in turn, comes in two varieties:

- **Global symmetry.** Acts equally everywhere. For instance, you add 100,000 volts to all points in space and the laws still apply.
- **Local symmetry.** Can vary with location or time. For instance, you can add a different voltage to every point and the laws still apply. Physicists also call a local symmetry a gauge symmetry. The term is a historical leftover and has nothing to do with oil gauges or 12-gauge shotguns. (Confusingly, gauge is sometimes taken to mean an internal symmetry. For clarity, I avoid the term.)

Global symmetry describes nature, and local symmetry describes our description of nature. Suppose I have a beach ball. It looks the same from every angle—that's the global symmetry. If you and I paint regular latitude and longitude lines on the ball, we can see its symmetry simply by rotating it until our views match. But why should what we see depend on how we draw the lines? If you paint drunkenly meandering lines and I paint nice regular ones, then to see that they're really equivalent, we can distort those lines by stretching and squeezing the surface of the ball, putting the plastic in tension. Each conceivable set of curvy lines corresponds to a certain amount of tension, and all are equally valid—that's the local symmetry, which can apply even when there's no global symmetry.

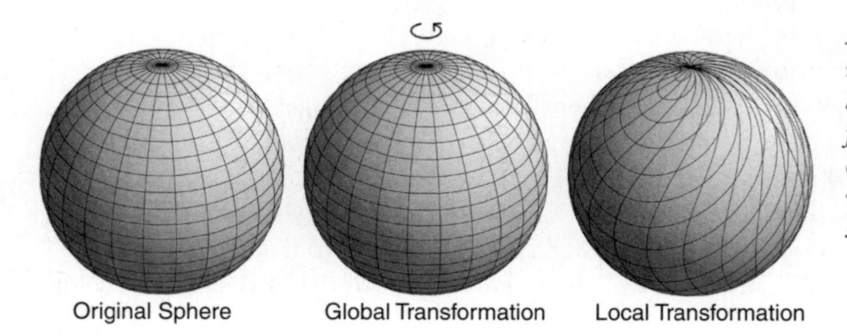

A global symmetry transformation rotates a sphere as a unit, whereas a local transformation moves each point on the sphere independently while maintaining its overall shape.

Ensuring local symmetry requires the laws of physics to be dynamic. A few subtleties aside, Einstein's theories exemplify the global versus local distinction. Special relativity is based on a global symmetry. If I'm sitting on a smoothly moving train and you're watching from the station platform, we can reconcile our perspectives with a straightforward offset in speed. It's just like rotating the ball by a certain angle to see that our grids line up.

General relativity is the local counterpart to this symmetry. If my train accelerates, we need to use different offsets at different moments in time. The analogue to the tension in the plastic of the beach ball is the artificial gravity pushing me back in my seat. According to Einstein's principle of equivalence, acceleration and gravity are two ways to describe the same situation. For the laws of motion to be the same for both of us, they must allow for the possibility of gravity.

In the case of electromagnetism, the global symmetry gives you conservation of electric charge and the local symmetry gives you the entire machinery of electromagnetic forces, photons and all. Those forces compensate for the fact that what you call zero volts, I could call 100,000 volts, a third person could call –5 volts, and we'd be equally right. All that really matters is the voltage difference at the point of contact, where the electromagnetic forces ensure that our conventions are consistent.

Quantum Leap

Because of local symmetry, if you want to supply 120 volts, you must provide a voltage reference against which to define it. That's the reason electricity requires two wires. The explanation we were all taught in elementary school, of completing a circuit, isn't quite right. Electrons don't flow around an AC circuit. They just oscillate back and forth, and they only need one wire for that. Besides defining zero volts, the second wire prevents the electric charge on an appliance from varying as electrons oscillate in and out.

The Standard Model is based on local internal symmetries. The symmetries that account for the nuclear forces are similar to the one for electromagnetism but somewhat more involved. What complicates them is that there's just one type of electric charge, but two types of weak nuclear charge and three quark colors. Those three quark colors, like ordinary colors, can mix. Physicists pick three as primary and call them red, green, and blue. But who's to say that what you perceive as red, I also perceive as red? The local symmetry of the Standard Model ensures that what you call a red quark, I could call mauve, and we'd be equally right. All that matters is the color contrast at the point where particles meet.

A tangible result of the nuclear symmetries is that they relate seemingly dissimilar particles and allow them to convert from one type to another. Quarks can don new colors. Particles with different weak charges, such as the electron and neutrino, can morph into each other by an operation akin to rotation.

Symmetries of Nature

Name	Type	What's Equivalent	What It Implies
Time translation	Spacetime	Moments of time	Energy is conserved
Space translation	Spacetime	Positions in space	Momentum is conserved
Rotation	Spacetime	Orientations in space	Angular momentum is conserved
Scale-invariance	Spacetime	Objects of different sizes	Forces hardly vary in strength with scale (see Chapter 6)
Special relativity	Spacetime	States of constant motion	Space and time are united
General relativity	Spacetime	All states of motion	Gravity is spacetime warping
Charge reversal (conjugation)	Spacetime	Particles and antiparticles	Antimatter exists
Mirror (parity)	Spacetime	Left and right	Particles come in left- and right-handed varieties

Name	Type	What's Equivalent	What It Implies
Time reversal	Spacetime	Moving forward and backward in time	Laws don't care which way time flows
Electro-magnetic gauge symmetry	Internal	Calibration of zero voltage	Electric charge is conserved (global); electromagnetic waves transmit forces (local)
Quark color symmetry	Internal	Choices of primary quark colors	Color is conserved (global); strong force binds quarks (local)
Particle exchange	Bridges spacetime and internal	Identical particles when swapped	Particles of matter resist packing (see Chapter 4)
Super-symmetry (hypothe-sized)	Bridges spacetime and internal	Particles of matter and particles of force	Every particle has a partner
T-duality (hypothe-sized)	Spacetime	Small and large distances	Spacetime is emergent (see Chapters 16 and 18)

Equally important, though, is how the Standard Model *isn't* symmetrical. The weak nuclear force is notoriously lopsided (see Chapter 5). If it were fully symmetrical, the weak force and electromagnetism would be two sides of a coin (the so-called electro-weak forces) rather than such distinct phenomena. Physicists think that, deep down, the weak force is a model of harmony. Early in cosmic history, its true symmetrical self was present for all to see. It was glorious; it was magnificent; it was barren—like one of Le Corbusier's ideally proportioned and utterly lifeless concrete plazas. For instance, because particles in the electroweak-symmetric universe lacked masses, they traveled at the speed of light and couldn't settle down into compact atoms.

For the universe to become fit for complex structures, the symmetry had to break. It happened when one of the elements of the Standard Model, the Higgs field, froze out and filled space. The electroweak forces split, and the weak force became unable to propagate freely. Many of the contingent features of our universe (see Chapter 15) date to this event. It was as if a fog settled on the concrete plaza, hid its harshness, and made it seem quirkily charming.

Who's the Most Symmetrical of Them All?

The symmetries of the Standard Model relate particles within classes: red quarks to blue and green ones, electrons to neutrinos, and so on. The putative grand unified theories break down these class barriers and relate all the particles of matter (quarks, electrons, neutrinos) to one another. But that leaves one great imbalance in nature. What, if anything, relates particles of matter to particles of force?

The answer, inspired by string theory, is *supersymmetry*. It performs a grand dynastic marriage of the particle families, stating that every fermion (the category that encompasses particles of matter) pairs off with a boson (the category that encompasses particles of force), and vice versa. All the particles of the Standard Model have a partner particle, or *sparticle*, waiting to be found. In anticipation, researchers have come up with a veritable Pig Latin for naming them. Quarks hook up with a new breed of particle known as squarks, leptons with sleptons, ups with sups, tops with stops, and so on. For bosons, the partner names end with "ino." Photons get together with photinos, W bosons with winos, and so on. Apart from enlarging the family of particles, supersymmetry enlarges the scope for making puns, as if *p*-branes and G-strings weren't enough.

def•i•ni•tion

Supersymmetry is the symmetry relating fermions (including particles of matter) and bosons (including particles of force) to each other.

A **sparticle** is a supersymmetric partner of an ordinary particle.

What makes supersymmetry so super is that it forges a link, not just between the particle categories but also between spacetime symmetries and internal symmetries. Using an abstract version of rotation, you can transform particles into sparticles and then back again. In the process, the particles scoot over a little bit in space. A change in an internal property affects an external one. Before they discovered supersymmetry, physicists had reckoned such a feat impossible.

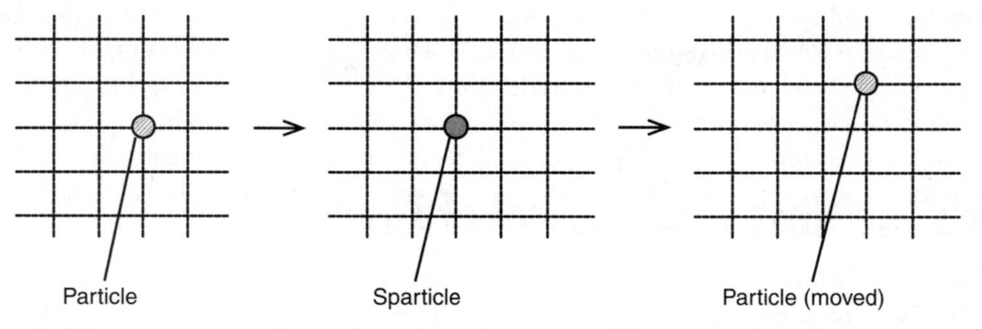

Applying a supersymmetric transformation twice causes particles to move through space.

Like other symmetries, supersymmetry comes in both a global and a local version. If every local symmetry has an associated force, what is supersymmetry's force? The fact supersymmetry causes a particle to scoot over suggests the answer. Motion is described by special relativity, and the locally symmetric version of special relativity is general relativity. It stands to reason that local supersymmetry is related to general relativity—and, therefore, to gravity. This is one of the most dramatic examples of how uniting the elements of quantum theory may rope in gravity as well.

In the 1970s, physicists sought to develop local supersymmetry into a quantum theory of gravity known as *supergravity*. Unfortunately, it scooted straight into a brick wall. For one thing, it suffered, like so many other efforts to quantize gravity, from the paradox of the graviton—the particle that transmits gravity would spiral out of control (see Chapter 11). Embedding supersymmetry in string theory solved this problem.

def•i•ni•tion

Supergravity is the theory of gravity based on supersymmetry.

Super Well Hidden

Supersymmetry is beautiful, pure, and ultimately as sterile as other symmetries. It must be broken because if it weren't broken, every particle and its sparticle would have the same mass, physicists would have seen them all by now, and supersymmetry would be an established fact rather than just a good idea. Seeking sparticles is one of the main goals of the Large Hadron Collider. In order to have eluded detection, sparticles must be heavier than ordinary particles and therefore beyond the capacity of older accelerators to create. In addition, they must not be totally free to decay into ordinary particles, or we'd have seen them in naturally occurring particle processes. The least massive sparticle, like the least massive ordinary particles (electron, up quark), must be stable.

Like the electroweak symmetry, supersymmetry had to break sometime early in cosmic history. How it happened, physicists aren't sure. They suspect it happened not long before the electroweak symmetry broke. In fact, the one may have triggered the other. For our own good, supersymmetry had to break in the right way. If, for example, selectrons had turned out to be lighter than electrons, atoms would consist of selectrons rather than electrons. Whereas electrons are standoffish and resist being shoved together, selectrons have no such compunction. They'd all pack into the same orbit close to the atomic nucleus, rather than space themselves out in orbits of different sizes. Two atoms, rather than repelling, would meld. Physicist Robert Cahn of Lawrence Berkeley National Laboratory, who has pondered this and other what-if cases in physics, compares the outcome to "Ice-Nine in Kurt Vonnegut's *Cat's Cradle*, only worse."

Pros and Cons

Even in its impure form, supersymmetry clears out many of the nettles of modern physics:

- **It explains why matter exists.** Whereas particles of force emerge organically from the requirements of local symmetry, particles of matter don't. In standard quantum theory, they have to be assumed from the outset. According to supersymmetry, the existence of force particles implies the existence of matter particles since they can be transformed into each other.
- **It solves the Higgs hierarchy problem.** As mentioned in Chapter 7, the Higgs field is a narcissist and interacts with itself intensely, emitting particles and then reabsorbing them in a juggling act that gets totally out of hand. The energy of its self-interaction acts as mass, elevating it far above its apparent value. Supersymmetry doesn't presume to deny the Higgs its pleasures but evens them out. For each particle the Higgs emits, it also gives off a sparticle, whose effect negates that of the particle.
- **It ameliorates the cosmological constant problem.** For similar reasons, supersymmetry brings some balance to the fluctuations that fill otherwise empty space with energy. It doesn't zero out the energy but does greatly reduce it, which is progress.
- **It causes the forces to converge in strength.** As mentioned in Chapter 6, particles are surrounded by entourages of virtual particles, which cause the forces of nature to vary in strength as physicists apply more energy. Eventually the forces

become nearly equal in strength—a sign that they are related. If virtual sparticles join in, they fine-tune the rate of variation. Electromagnetism strengthens with energy faster, the strong force weakens with energy slower, and the weak force shifts from weakening to strengthening. The end result is that the forces don't just become nearly equal in strength but almost *exactly* equal.

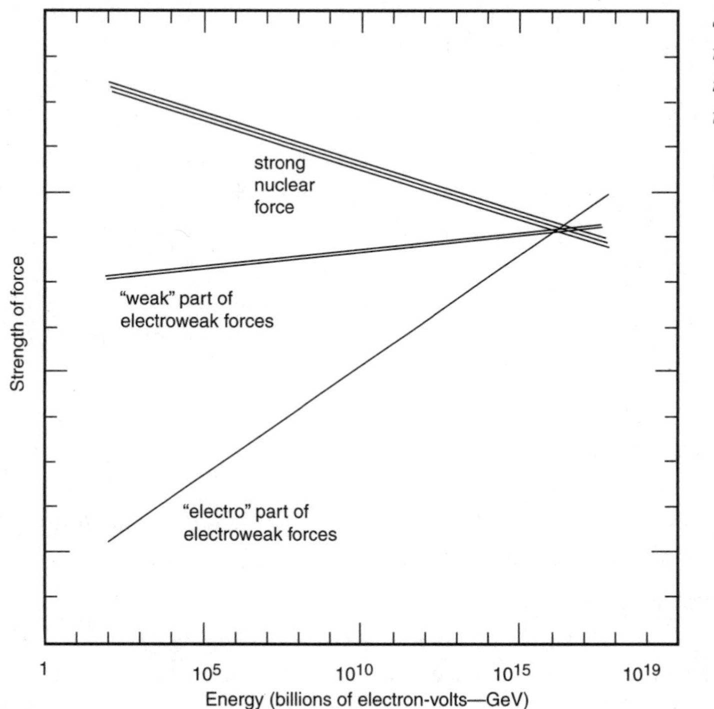

Supersymmetry changes the way the forces of nature intensify with energy, so that they meet up exactly.

(Courtesy of Keith Dienes)

- **It stabilizes the proton.** Grand unified theories used to predict that the proton should decay at a slow but discernible rate. By pushing up the energy at which the forces converge, supersymmetry can slow its decay, which would explain why physicists have yet to see a proton fall apart.
- **It may explain the dearth of antimatter.** The near-total lack of antimatter in our universe has long been a mystery since matter and antimatter behave so similarly. For particle reactions in the early universe to have favored matter to such a degree, they must have been thrown badly off-kilter (see Chapter 7). Supersymmetry could have done that.

- **It may explain the universe's dark matter.** Known particles can't account for the dark matter that fills our universe. Sparticles can. Processes early in the big bang would have created them as surely as they created ordinary particles. The lightest, being stable, would linger to this day in amounts that match astronomers' observations. Physicists think this sparticle is the so-called *neutralino*, a cocktail of the photino, higgsino, and zino. As its name suggests, the neutralino is electrically neutral, so it doesn't interact with photons. It would indeed be dark.

def•i•ni•tion

The **neutralino** is thought to be the lightest possible supersymmetric particle.

That said, supersymmetry raises some questions that physicists have yet to solve. Sparticles haven't been observed; the process that broke the symmetry remains a mystery; and novel supersymmetric phenomena haven't been seen. It's possible that supersymmetry broke at such a high energy that it doesn't bear on the problem of the Higgs or the cosmological constant, in which case its charms are more theoretical than practical.

A Higher Point of View

Besides introducing physicists to supersymmetry and other new symmetries, string theory has encouraged physicists to rethink the concept of symmetry. For instance, strings blur the categories of spacetime symmetry and internal symmetry. If space has extra dimensions we can't see, then many of the internal symmetries of particles could actually be spacetime symmetries. They look internal only because the extra-dimensional space is hidden from us. The seemingly abstract rotations of the internal symmetries could be real rotations, but in higher dimensions. In fact, that's why string theory predicts space has a total of 10 dimensions. If it had more, there'd be too much symmetry. No laws of physics could satisfy its strictures.

In addition, string theory shows that many symmetries *aren't* as essential as physicists once thought. Physicists long suspected that the specific symmetries of the Standard Model weren't truly fundamental but, instead, were aspects of the more encompassing symmetry of a grand unified theory. Nonetheless, they thought the concept of local internal symmetry would endure.

String theory calls that into question. The concept of duality, introduced in Chapter 16, is a symmetry of symmetries. If symmetry suggests that a seemingly fixed aspect of the world can be an artifact of our point of view, then duality suggests that symmetry itself can be an artifact of our point of view. Different theories of the world, involving different symmetries, can be completely equivalent. If swapping one symmetry for another has no real effect, then the symmetry can't be fundamental. It must be more of a tool than an objective fact. In a sense, that's what physicists approaching quantum gravity from the metaphor of fluids and solids have been arguing all along.

In the Loop

Gauge symmetries might not be fundamental! They might only appear as long distance artifacts of our description of the theory. If so, perhaps some of the gauge symmetries of the standard model or even general relativity are similarly long distance artifacts.

—Nathan Seiberg, Institute for Advanced Study

If the symmetries physicists have been working with aren't fundamental after all, what is? No one yet knows. In the meantime, string theorists happily use duality to switch between symmetries and explore their theory from different angles. So the pursuit of beauty may be taking physicists beyond conventional notions of it.

The Least You Need to Know

- Modern theories of physics are built on the idea of symmetry, and string theory pushes this idea to its logical conclusion.
- The concept of supersymmetry, which was inspired by string theory, unifies particles of matter and of force.
- Supersymmetry ties up a huge number of loose ends in physics, possibly cosmic dark matter.
- String theory implies that many of the symmetries assumed in relativity theory and the Standard Model of particles may not be truly fundamental.

Chapter 18

Emergence

In This Chapter

- Spooky action at a distance
- Is the universe a living hologram?
- Making space where none was before
- Pulling threads from the spacetime fabric

This chapter focuses on the idea that space and time, among the most fundamental things you can think of, are actually the product of something still simpler—some basic ingredients that rise above concepts of place and moment. Those ingredients play off one another and produce something that wasn't there at the outset.

Emerging Ideas

One of the most annoying phrases in our language is "Make time for it." If only we could! Short of creating a black hole, inside of which a new region of spacetime might bud off, we're stuck with the time we have. All we can do is rob it from Peter to linger on the phone with Paul. If bosses, teachers, and parking attendants recognized these trade-offs, the world would be a happier place.

For quantum-gravity theorists, though, making time is a rallying cry. Their ultimate goal is to figure out where the notion of spacetime comes from. Is it a foundational concept, incapable of further explanation, or is it emergent, assembled from something still deeper? It's hard even to imagine what that something could be. How can we think of things when there's no place they could be? How do they assemble if assembly is a process occurring in time?

Yet the task is no different in kind from how any phenomenon emerges from what lies below it. Quarks interact to produce protons, protons to produce atomic nuclei, atoms to produce molecules, and so on until we get creatures that want it all done yesterday. Along the way, new qualities emerge that weren't obvious in the original building blocks. So it may be with space and time.

The candidate quantum theories of gravity are all very far from achieving this goal. Every effort to step outside of space and time presumes some aspect of it; it's hard to break free altogether. But physicists have loosened a few of the restraints. The slipperiness of the concept of distance, discussed in Chapter 16, hints at how spacetime might emerge from shenanigans in the microscopic realm. Strings seem inseparable from the space they reside "in." If we squeeze down a cylinder, the string residing on it can grow as though it makes room for itself. The way the string winds around the cylinder is indistinguishable from motion in a new dimension of space, one that isn't explicitly included in the equations but materializes of its own accord. Small space, large space—to a string, it's six of one and half a dozen of the other. If space were foundational, the concept of distance wouldn't be so flighty.

In the Loop

One does not really need spacetime anymore; one just needs a two-dimensional field theory describing the propagation of strings. And perhaps more fatefully still, one does not have spacetime any more, except to the extent that one can extract it from a two-dimensional field theory. … 'spacetime' seems destined to turn out to be only an approximate, derived notion.

—Edward Witten, Institute for Advanced Study

In loop gravity and buckyspace, too, microscopic spacetime looks nothing like the macroscopic sort. In these theories, the spacetime we observe is a quantum average of possible geometries on fine scales. How our macroscopic spacetime emerges is still an open question, but buckyspace has found an important condition: the four dimensions of space and time emerge from primitive tidbits of spacetime only if the tidbits are forced to respect cause-effect relations. Ensuring that cause and effect remain distinct seems part of the conceptual underpinnings of spacetime.

Reach Out and Touch Someone

In the traditional view of space, one of its essential properties is what physicists call locality. Locality means that each point in space and time is an individual with an independent existence. Points directly affect only the points they directly touch. For something to move from one place to another, it needs to cross through every point in between, which unavoidably takes time. It's fun to think about leapfrogging across spacetime using psychic powers or hyperspace jumps, but we must be careful what we wish for. Locality ensures that space is our protective moat. Violent events transpire all over the universe, and the intervening abyss attenuates their effects.

Without locality, the world would go haywire. Yelling at your TV set really might change the outcome of the game. Space invaders might flit across the galaxy, bop you on the head, and flit back before you know what hit you (see Chapter 3). Depending on your own speed, an instantaneous influence could look like one going back in time, so cause and effect would get muddled.

General relativity and modern quantum theory have elements of nonlocality, but the effects are contained. Quantum theory is widely regarded as nonlocal (see Chapter 4). Two particles can remain blood brothers however far apart they are. Conversely, two neighboring particles that lack this tight bond might be oblivious to each other. Spatial position hardly matters at all. Einstein called it "spooky action at a distance." Yet the particles' relationship is private; space invaders couldn't use it for their own nefarious ends.

As for general relativity, the malleability of spacetime makes individual points impossible to identify definitively. When we try to make observations at specific points, we run into conundrums such as the problem of frozen time (see Chapter 7). Measurable quantities are nonlocal; they can't describe points, only regions. Fortunately, this nonlocality, like the quantum sort, doesn't expose our planet to space invaders. Gravity is typically so weak that spacetime has a nearly fixed shape and provides a perfectly adequate scaffolding for pinpointing events. Another threat to locality in general relativity is wormholes, but in the absence of special quantum effects that might well prove impossible, wormholes collapse of their own accord (see Chapter 10).

As these two theories collide in a quantum theory of gravity, nonlocality intrudes more forcefully. All the leading approaches to quantum gravity involve some degree of nonlocality. Distance ceases to matter when it's short enough; we can't even define it anymore. The fraying of locality, such a basic property of space, suggests that space itself unravels. When something is nonlocal, it has no position; it somehow sits outside space. And if you want to explain what space really is, that's precisely what you want: something that is beyond it.

The Holographic Principle

It's not surprising that locality breaks down on very fine scales. Things are weird down there; this is just one more instance of weirdness to add to the list. What's more startling is that locality could also give way on large scales. To see why that might happen, let's consider a thought experiment involving something we don't normally associate with quantum gravity: computer memory chips. Today's chips use nanometer-scale transistors to store data bits. As engineers continue to shrink chips down, they might use atoms or even subatomic particles and make them three-dimensional rather than flat silicon wafers. Ultimately, engineers of the far future might exploit strings or whatever other building blocks are the tiniest possible.

If locality holds, the total information capacity is proportional to the volume of material. The whole is just the sum of its parts. If you visit the computer superstore of the future, a salesperson, salescyborg, or, depending on how optimistic you are about humanity's future, salescockroach would sell you eight string-based memory cards, advising you to slot them into your computer for eight times as much data as a single card.

But that doesn't give gravity its due. Each data bit corresponds to a certain amount of energy stored in the chip, so a lot of bits entails a lot of energy, which generates a strong gravitational field. If you try to pack too much data into the chip, gravity becomes strong enough to create a black hole (giving new meaning to a computer crash). If you try to add capacity by stacking eight cards, arrayed in a cube two on a side, the force of gravity intensifies, so you can store only four times as much data before triggering the black hole. Each card can hold half as much data as it did on its own. The total capacity goes up with the surface area rather than the volume.

In short, when gravity is strong, the whole is less than the sum of its parts, which goes against what we expect from the concept of locality. This violation of locality can occur on any scale; it is not confined to the microscopic.

Down the Memory Hole

The fact that information storage capacity is proportional to area rather than volume is perplexing and profound, so let's look at it again in more detail. It all has to do with black holes. The processes that govern them set the limit to the information content of more mundane objects.

Black holes, like anything else, can store information. We know this because they not only suck in matter but also glow like a hot coal (see Chapter 8). For an ordinary object, heat is associated with molecular motion, and although physicists don't know what a black hole's "molecules" (its basic constituents) are, the mere existence of these molecules implies that the hole encodes information. The number of gigabytes in a black hole depends on the number of possible molecular arrangements, a quantity known as *entropy*.

def•i•ni•tion

Entropy is a measure of the number of possible ways molecules in a substance can be arranged for a given amount of energy.

To calculate the amount of entropy, physicists surmise the molecular properties from how the black hole radiates. When a black hole devours matter and increases in mass, it increases in radius yet *decreases* in temperature. A bigger hole is heavier but cooler—the gravitational forces on the surface, which determine the temperature, are weaker. Temperature is essentially the average energy per molecule. As a hole grows, its total energy goes up while the average energy of its putative molecules goes down.

Evidently, as a hole grows, its total energy is being spread out over a larger number of molecules, leaving each with less. It turns out that the number of molecules must scale up with radius squared—that is, with area. The number of molecules, in turn, determines the information storage capacity of the hole.

Quantum Leap

Here's a little more of the mathematical reasoning. As a rule of thumb, both energy and entropy increase with the temperature of an object, but entropy increases one notch more slowly. When energy increases with the fourth power of temperature, entropy increases with only the third power. When energy increases with the first power of temperature, entropy increases with (approximately) the zero power. In a black hole, energy increases with the –1 power of temperature, so entropy must increase with the –2 power. Converting temperature to radius, entropy must increase with the second power of radius—that is, with area.

If area limits the information content of a black hole, it must limit the information content of anything, since anything can be crunched into a black hole. By generalizing the definition of area, the limit applies even to expanding objects, which don't have a fixed area in space. That includes our universe as a whole.

Quantitatively, each Planck area (square Planck length) accounts for about one bit of data. Those mysterious black-hole molecules, whatever they may be, must be Planckian in scale, which is further confirmation that quantum gravity governs them. One bit per Planck area is an awful lot of information for any decent-sized body. Even if we tallied up the properties, positions, and velocities of every particle in an ordinary body, it wouldn't come close to maxing out this limit.

Even so, the limit is a huge comedown from what locality implies. Suppose that standard quantum theory, which assumes locality, says an object is so complicated that we'd have to commandeer the entire Internet (about an exabyte of data) to store its blueprint. The area information limit says a single 100 gigabyte hard drive would do. That's a factor of 10 million less. For larger objects, the factor is even greater. Our universe as a whole can store roughly a googol (10^{100}) bits, a factor of about 10^{60} less than locality would imply.

Adventures on the Holodeck

Because the assumption of locality misses the mark by a huge factor, the points that make up spacetime must not be independent even in principle. They're interconnected by a cat's cradle of gravitational forces, so that information content depends on area rather than volume. You could store everything there is to know about an object on its surface. Theoretically, not only can you judge a book by its cover, but you can also read it by its cover. The book jacket carries a subtle imprint of every word.

def•i•ni•tion

A **hologram** is a special type of photograph that captures an entire 3-D scene so that it can be viewed from various angles.

The **holographic principle** holds that the amount of information in a region is proportional not to its volume but to the area of its boundary.

Physicists call this concept the *holographic principle*, by analogy to a *hologram*, which stores a full 3-D scene on a flat sheet of film. The trick of a hologram is nonlocality. It stores the image not as splotches of pigment corresponding to individual objects in the scene but as a wave pattern, which spreads out each object across the whole film so they're all overlapping. When you illuminate the hologram, you reconstruct the light waves. A third dimension that isn't really there emerges from the way the information is stored on the hologram.

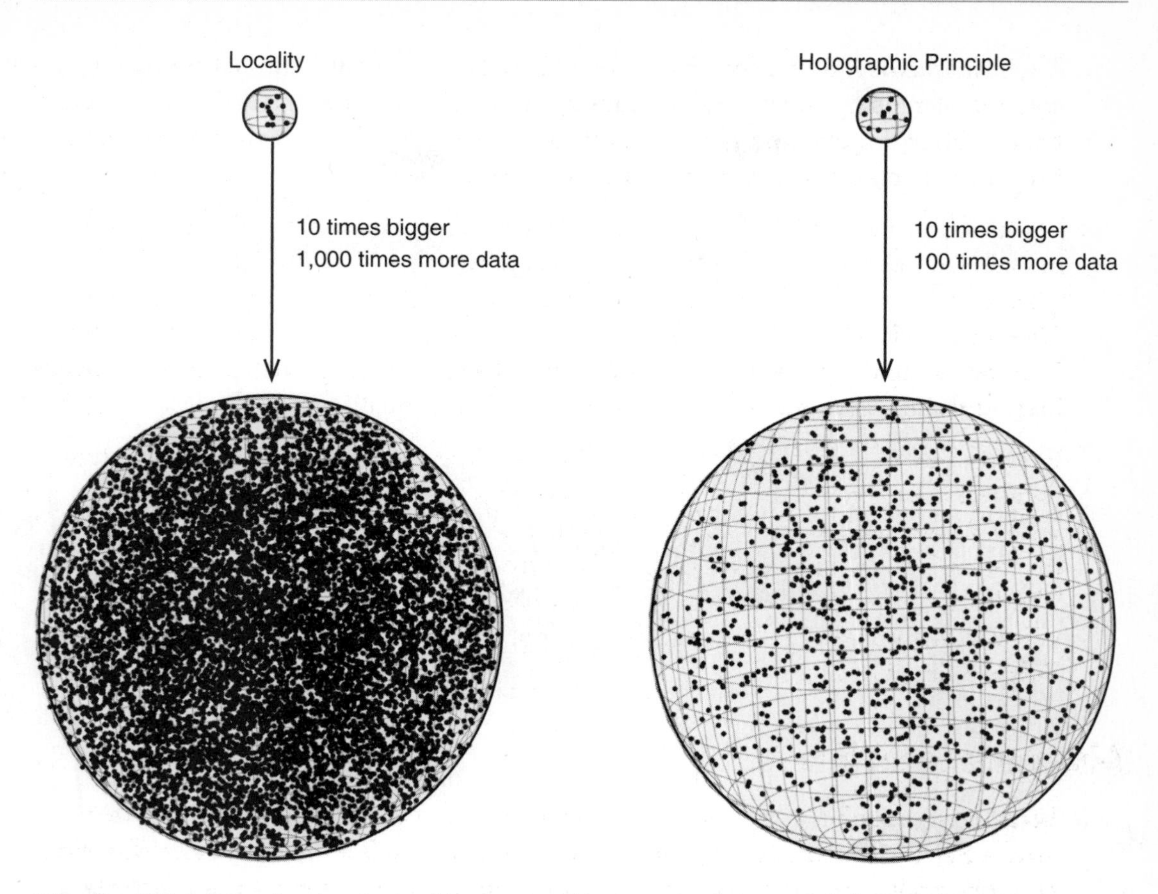

If locality held, information capacity would scale up with volume, but because of the holographic principle, it actually scales up with area.

Similarly, one of the dimensions in space may not be fundamental but could emerge from what happens in a lower-dimensional space. What we perceive as widely separated objects may fall right on top of one another in that space, and something besides spatial separation keeps them from interacting.

Stringy Holography

Some married couples can sit at the same table with each other but might as well be in different time zones. Lovey-dovey couples can be in different time zones, yet feel they could touch each other. Proximity is a matter of perspective. This maxim underlies the most fully developed account of what holography tells us about space and time.

The idea is to exploit the concept of duality, the fact that strings can behave in two different yet completely equivalent ways. Duality has become string theorists' Gerber multi-tool, packed with so many functions they're still figuring it out. Using it, theorists have shown that almost everything that seems fundamental isn't: distance (see Chapter 16), symmetry (see Chapter 17), and now spacetime itself. Duality can match up entire universes so that whatever happens in one has a counterpart in the other. Those universes can have different numbers of dimensions, just as a flat hologram is equivalent to a 3-D scene.

The classic case, put forward in 1997 by Juan Maldacena, now at the Institute for Advanced Study, in what became the single most influential theoretical physics paper of the '90s, pairs a universe with five dimensions of space and time and a 4-D universe that forms its outer boundary, like an apple and its skin. The 5-D interior universe is a normal one, at least by string-theory standards. It's filled with strings that exert gravitational forces on one another. The 4-D boundary universe is filled with quantum particles that don't exert gravity. These particles are essentially the Standard Model on steroids. Instead of one, two, or three forms of charge (such as electric charge or quark color), the particles can have billions or more. This extra complexity makes up for the lack of gravity and ensures that the two universes behave in the same way. The larger the interior spacetime, the more types of charge the boundary universe needs.

def•i•ni•tion

Gravity/gauge duality is the equivalence of a universe with gravity to a lower-dimensional universe without gravity.

Because the interior has gravity and the boundary has only quantum forces, which are based on so-called gauge symmetries (see Chapter 17), the duality between the two universes is known as *gravity/gauge duality*. Since 1997, string theorists have found other universes that match up with each other, so they suspect it's a general principle. To be fair, though, they have yet to prove it, and some physicists are still skeptical.

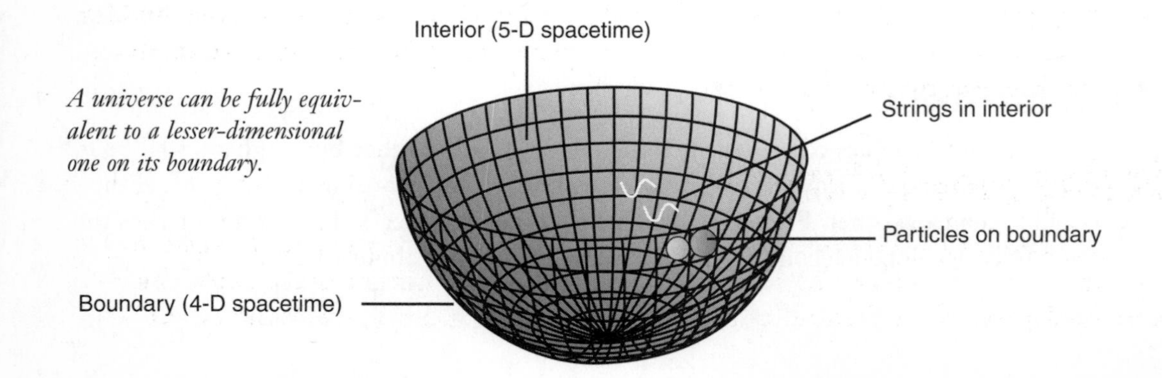

A universe can be fully equivalent to a lesser-dimensional one on its boundary.

Although the two universes are technically equivalent, string theorists commonly think of the boundary universe as primary. A region in the interior has a limited amount of information because it's governed by processes on the boundary, which has one dimension fewer. The boundary processes themselves are purely local. No gravity gets in the way because gravity is itself derivative.

The duality also hints at how space can arise from spacelessness. In Chapter 6, I explained how processes occurring at different size scales are almost independent of one another. Each scale directly affects only the next bigger one. The situation is like the parable of the kingdom and the nail—the nail affected the horseshoe, which affected the horse, then the rider, then the battle, then the kingdom. For an influence to move between widely separated scales, it must cross through every scale in between.

Quantum Leap

Physicists working on loop gravity and causal sets have a different take on the holographic principle. They think of it as a limit not on the information content of space but on how much of that information can get through to us when we conduct observations.

This independence of scales looks just like the independence of points in space. The gravity/gauge duality takes this resemblance literally. Two particles residing at the same location in the boundary universe barely interact if they have different energies. Their counterparts are strings at two different locations inside the interior universe. Particles could rub shoulders on the boundary, while their stringy avatars are light-years apart in the interior. Distance may be a grand illusion, which perhaps offers some consolation to the separated lovers of the world.

Although just one dimension emerges, gravity/gauge duality has become the template for understanding in general how space and even time might emerge. For instance, quantum particles might initially interact in a way that, like a badly lit hologram, doesn't give rise to an interior universe. Gradually, though, they start behaving in a way that mimics the additional dimension. Space or time arises, and you get the same "ooooooh" sensation as when you first see a hologram pop out at you. Even this idea, though, presumes that time operates on the boundary. String theorists have yet to create space and time completely from scratch.

Seeing Spooky Action?

No one has yet detected any nonlocal transmission of information. In the right situation, though, locality might get ratty and let us catch a glimpse of the underlying nonlocality. Some theorists think strings that are in contact on the boundary, yet far apart in the interior, might be able to communicate with one another, perhaps explaining the mysteries of black holes (see Chapter 19) or even the nonlocality of standard quantum theory.

In the leading alternative theories—loop gravity, buckyspace, and causal sets—points in the microscopic spacetime connect only to their nearest neighbors. But points that are neighbors on the connect-the-dots figures of these theories may be far apart in spacetime, like parts of a sewing pattern that are next to each other on the page yet far apart on the body. If you don't know what you're doing, you might end up with threads connecting the cuff to the collar. Lee Smolin and Fotini Markopoulou of the Perimeter Institute call this disordered locality. Such straggly connections are presumably very rare, but space is big, so the absolute number of them could be enormous. They act like miniature wormholes and might allow forces to leak from one part of space to another, with subtle effects on cosmology (see Chapter 20) and quantum theory (see Chapter 23).

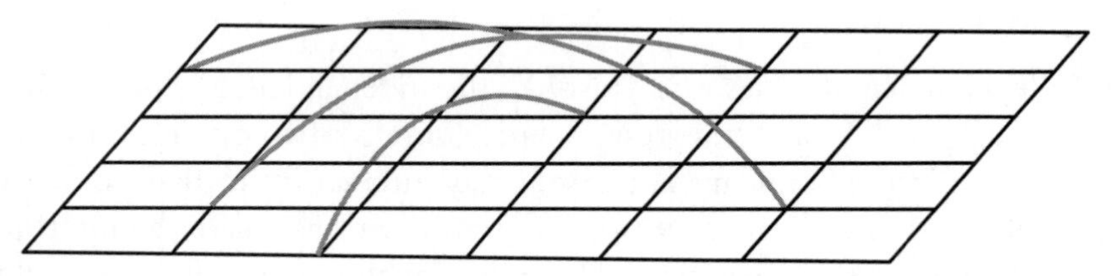

Like missewn threads between different parts of a shirt, nonlocal links might connect far-flung points in spacetime.

All the features of candidate quantum gravity theories discussed in this part of the book—extra dimensions, multiple universes, foamy spacetime, symmetries of seemingly different particles types, and nonlocality—have a trace of magic about them. Some people might misinterpret this to mean that physics proves the existence of miracles. In fact, these features are well hidden from us and have to be for our own good health. What makes the everyday world magical, a place just right for us, is not that we could stick our arm into an extra dimension or leap across space with a single bound. It's that we *can't*.

The Least You Need to Know

- Space and time may be derivative concepts.
- The so-called holographic principle is one piece of evidence for this.

- In string theory, the holographic principle reflects a connection known as gravity/gauge duality, which allows space to emerge from spacelessness.
- Space may be riddled with nonlocal connections that tie together far-flung points.
- These connections may become apparent in extreme situations such as black holes.

Part 6

What Has String Theory Done for You Lately?

Having studied the core ideas of string theory and its cohort, let's apply them to the knotty problems we looked at in Part 3, such as what happens inside black holes, what came before the big bang, and why quantum theory is so inscrutable. Although getting hard evidence is, well, hard, experimenters are already nibbling around the edges of the theories. Many scientists are dissatisfied with the rate of progress, but they haven't come up with any viable alternatives.

Chapter 19

Black Branes and Balls of String

In This Chapter

- Black holes as a new state of matter
- Where'd the information go?
- Loopy black holes
- Back to time machines

The saying goes that in times of stress, you can see what people are really made of. Much the same is true of matter and spacetime. No situation is more stressful than a black hole. The two leading candidate quantum theories of gravity offer deep, if still patchy, explanations of these cosmic sinkholes. In string theory, a black hole brings out the true stringiness of matter; in loop gravity, it reveals the atomic nature of spacetime. As for the related phenomenon of time machines, the theories send mixed messages.

Getting Warm

If the goal of physics is to unite all the scattered children of nature into one big family, the prospective family member that'll require the most coaxing is the black hole. Leaving aside its destructive tendencies, the black hole just seems too different from protons and photons to ever be brought under the same conceptual roof. So the challenge of a quantum theory of gravity is to domesticate this beast.

As established in Chapter 8, black holes raise two questions. First, what exactly happens to matter that falls into one? Einstein's general theory of relativity says it's compressed into a mathematical point (the singularity) at the center of the hole, achieving infinite density. An infinite anything is scarcely believable. Second, what happens to the information embodied in the matter? Is it lost or somehow stored for later retrieval? Loss would violate quantum theory, and storage would make the black hole a peculiarly methodical pack rat.

String theory treats both problems as aspects of a deeper issue. According to general relativity, a black hole converts matter into pure gravitation—that is, into the warping of space and time. For matter and spacetime to interconvert, they must really be made of the same stuff. So black holes are not just giant monsters out in the cosmos; they are actually windows into the ultimate composition of matter.

Suppose you have an ice cube and start heating it up. It melts to water, then boils to steam. If you keep the heat on, the H_2O molecules dissociate into atoms, the atoms into protons and neutrons, and the protons and neutrons into quarks. You reach billions and trillions and billions of trillions of degrees and eventually the particles begin behaving as strings.

Finally you reach a point where despite pumping energy into the strings, you can't raise their temperature. They keep finding new ways to vibrate, so the average energy per vibrational pattern (which is what temperature represents) stays fixed. This temperature is known as the *Hagedorn temperature* of strings, which is equal to about 10^{31} kelvins.

def•i•ni•tion

The **Hagedorn temperature,** in string theory, is either the maximum temperature that strings can attain or the point of transition to a new phase of matter.

A fixed temperature sounds like what happens when you melt ice. As you heat the ice, its temperature goes up, but when it reaches the melting point, the temperature plateaus. The heat you apply goes into changing the phase from solid to liquid. Only when the transition is done does the temperature resume its upward march.

By analogy, strings appear to "melt" at the Hagedorn temperature into a new form of matter. Intriguingly, this phase of transition occurs just when strings have so much energy that they can collapse into a black hole. (The one proviso is that the strings must also be coiled up.) So maybe the black hole represents a new form of matter. At first glance, black holes and strings are mismatched. The information storage capacity of a string is proportional to its length and, therefore, to its energy. The information storage capacity of a black hole is proportional to its area and, therefore, to its energy squared (see Chapter 18). In addition, a string's temperature rises with energy, whereas a hole's temperature decreases with energy.

Amazingly, though, the information capacities and temperature values are numerically equal at the melting point. So strings do match up with black holes after all. A hot string melts into a hole or, going the other way, a hole freezes to form a string. As a black hole emits radiation, its energy diminishes until it turns into strings, which cool off and turn back into recognizable particles, maybe even an ice cube. For many physicists, this clean dovetailing of strings and black holes alone justifies their interest in string theory.

Quantum Leap

String theorists aren't the only ones who conceive of black holes as transitions between phases of matter. Physicists approaching quantum gravity from the inspiration of ordinary fluids and solids do so as well. They conjecture that relativity theory itself fails at the transition, much as the laws of fluid flow seize up when water freezes. So far, though, the idea is more of a hazy analogy than a solid theory. A related idea is that black holes are gravastars, where gravity switches from being an attractive force to being a repulsive one. The repulsive interior of the gravastar counterbalances the inward pressure of the overlying material. The theory hasn't attracted much support, though, because it doesn't explain how gravity could make the transition.

Melting Pot

This matching argument applies at the level of an individual string, whereas a black hole really involves the collective behavior of gazillions of strings. To capture them en masse, string theorists whip out their favorite conceptual tool: duality, which holds that two situations that seem different are actually equivalent. In the case of the so-called gravity/gauge duality discussed in Chapter 18, a scenario where gravity operates is equivalent to one where it doesn't but where particles engage in a complex quantum choreography. The two scenarios don't look anything like each other; they don't even operate in the same number of space dimensions. But they give identical answers to equivalent observations.

So this duality lets string theorists do some lateral thinking. The particle choreography may be complex, but it's still easier to follow than the workings of gravity, so we can start off with a gravitational situation, translate it into a nongravitational one, and then sort out what happens. Everything in one scenario has a counterpart in the other. A recognizable object such as an ice cube in the gravitational scenario corresponds to a clump of particles in the nongravitational one. Heating the object corresponds to heating the clump.

Eventually the object gets so hot that it collapses into a black hole, and at that moment the clump breaks up into individual particles. It literally melts. The resulting fluid, like any fluid, has a temperature, which matches the temperature calculated on the gravitational side. Because the particles obey quantum theory, they don't lose any information when they turn from clump to fluid. From this we can conclude that the black hole they correspond to doesn't lose any information either.

A body falling into the hole corresponds to a clump of particles plopping into the fluid and dissolving like a sugar cube in hot tea. The fluid retains a memory of the clump. The radiation emitted by the hole must not be purely random but must subtly encode the complete blueprint of everything that ever fell into the hole.

That, in itself, is a major discovery. It changes what physicists think would happen if you jumped into a black hole. I still feel compelled to advise you against it. But at least you wouldn't be completely lost—just thoroughly mashed. The information encoded in your body would outlive you, and the radiation emitted by the hole would subtly convey your state of mind and health at the moment of your demise.

Quantum Leap

Another interesting case study is the Flatland black hole—namely, a black hole in a 2-D space. It's a little surprising that such a hole is possible at all. A planar space lacks the complexity you'd think might be essential for one. In particular, objects in 2-D can't attract one another gravitationally. Even so, if space has the right shape, objects can get stuck in a region just as if they were in a higher-dimensional black hole. Applying gravity/gauge duality, physicists have confirmed these holes have an information-storage capacity that scales up with their area, which in two dimensions is the circumference of the hole. These holes even emit radiation.

A Black Hole Built of Branes

But what exactly is the new phase of matter in the hole? How exactly does the radiation it gives off reflect what fell in? How does information go the wrong way on what was supposed to be a gravitational one-way street? Ah, those are the big questions! Answering them requires translating from those complex particle interactions back to gravitational phenomena, which is like translating from one language to another without the benefit of a dictionary. You have to start with simple things you can point to ("Beer. *Cerveza*. Beer. *Cerveza*.") and slowly move up from there. Theorists have focused on special cases that are computationally easy, such as holes that have ceased emitting radiation because they have a charge like an electric charge that stabilizes them. Such holes can't actually exist in our universe, but they serve as a proof of principle.

Within our visible three dimensions of space, a black hole looks bland and featureless, but in the hidden extra dimensions lurks a swarm of branes. Strings stretch from brane to brane in a big tangle not unlike the wires behind my wife's desk, which she freely admits is something of a black hole. These strings are the molecules of the black hole and encode its information. The number of their possible arrangements is proportional to the area of the hole, in accordance with the holographic principle discussed in Chapter 18.

Some theorists argue that, when you work through what the brane swarm is doing, the singularity at the core of the hole isn't a pinprick after all—but quite the contrary. The strings stretching between branes in the extra dimensions vibrate in the lateral direction, which is to say, they vibrate in the visible dimensions of space. The strings have so much energy that their oscillations swell up and fill out the space within the hole's boundary, which can be millions of kilometers across. Being so large and floppy, the strings get tangled like a ball of yarn or a fuzzball, as Samir Mathur of The Ohio State University calls it.

The boundary of the fuzzball is not, in fact, a point of no return for infalling matter. As you fall through the boundary, you scarcely notice anything unusual, since the fuzz is so diffuse. You and other material pile up near the center of the hole and then the fuzz gets to work, picking you apart and assimilating you. It takes eons. The fuzz is ever so slightly changed for your presence. The radiation it gives off carries your imprint. What happens is much the same as if you fell into Jupiter. You'd gradually disintegrate, alter the structure of Jupiter a (very) little bit, and skew the radiation the giant planet gives off.

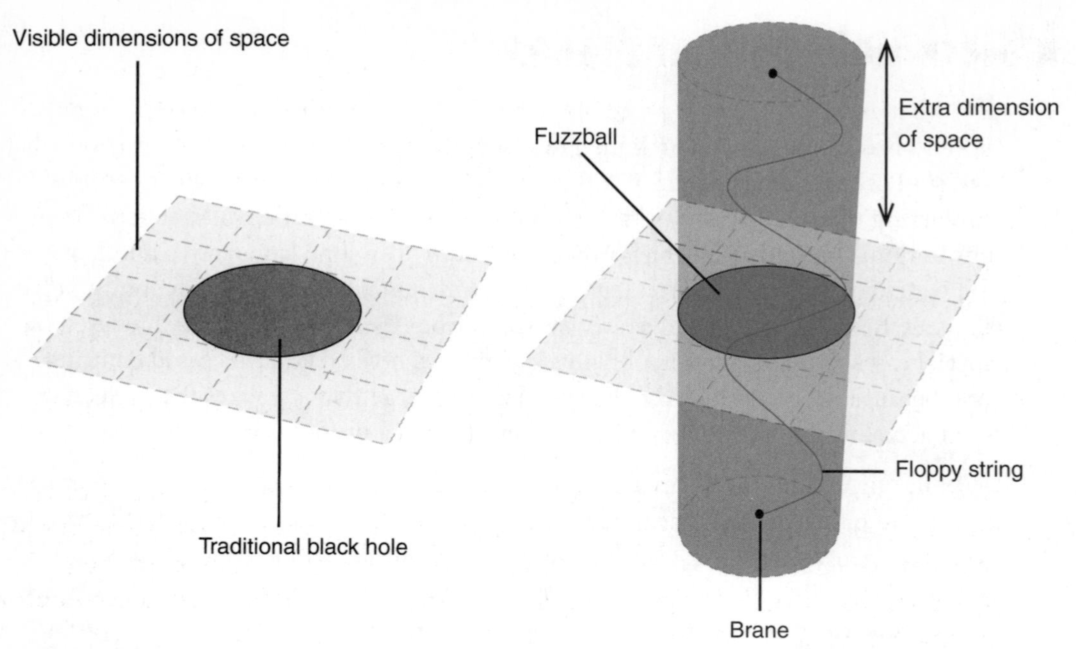

What looks like a black hole in our 3-D universe (depicted here as 2-D for simplicity) may have some behind-the-scenes activity in the extra dimensions of space.

The fuzzball concept remains fuzzy. String theorists can describe black holes as static objects but haven't been able to follow how they develop over time. One thing is fairly clear, though. The fuzzball and other stringy explanations involve the breakdown of one of the basic features of spacetime as we know it: locality. Points remain linked even though they appear to be too far apart to communicate. For the fuzzball, strings interconnect points within the ball. After something falls in and gets assimilated, its position becomes undefined. This might be our glimpse into what underlies space and time.

Quantum Leap

Regardless of what black holes actually are, string theory might alter them in one other respect. To make a black hole, you need to squeeze matter into a small enough volume. Gravity isn't strong enough to do so if the mass is less than 10^{-8} kilogram, the mass equivalent of the Planck energy scale. But if gravity is diluted by extra dimensions, it might be stronger than it seems. The smaller the distance, the less dilution has occurred and the stronger gravity might be. If so, holes lighter than 10^{-8} kilogram might exist, and particle accelerators might be able to make them (see Chapter 21).

Loop Hole

Unlike string theory, loop gravity doesn't associate black holes with a new phase of matter, at least not directly. They're still contorted regions of space with much the same geography as they have in general relativity: a singularity surrounded by a boundary marking the gravitational point of no return. But loop gravity smoothes out the rough edges of this description.

The boundary never settles into a perfectly smooth, round shape. It fluctuates and oscillates at the quantum level. These ripples file away the information of infalling bodies and imprint it onto the outgoing radiation. Areas in loop gravity are naturally divided up in units of the Planck scale, which is just what is needed to explain the hole's information capacity.

The singularity never achieves infinite density because the atomic nature of spacetime puts a cap on how much energy you can squeeze into a region. The wavelengths of photons can be no smaller than the Planck scale. If you try to push more energy into space, space pushes back—quite literally. Gravity switches from an attractive force to a repulsive force.

This might explain not only black holes but also their troubled siblings, naked singularities (see Chapter 8). In these black holes without boundaries, we can see all the way down to the ultradense wad of matter at their core. However, physicists debate whether such things can even get close to forming. Long before quantum effects kick in, ordinary gravity should warp spacetime to demarcate a point of no return. But if it somehow failed to do so, loopy effects would step in and cause the wad to blow apart in a grand explosion.

The repulsive force also could cause space to expand within a black hole, giving birth to a baby universe. It's like a parallel universe, except that it's not parallel so much as nested within our universe like a Matryoshka doll. The boundary of the hole hides the newborn universe from us. The baby universe endures even when the black hole appears to have evaporated away. Evaporation simply means that the black hole cuts its umbilical cord to our universe. Material from our universe can no longer flow into it, but the baby lives on. For some theorists, a baby universe explains what happens to the information that falls into a black hole. It's never truly lost but takes up residence in the baby.

Both the loopy and stringy approaches to black holes have been criticized as incomplete, even contrived. A more charitable way to put it is that they've done better on the question of whether black holes lose information—the answer: probably not—than on where the information gets stored and how it gets back out again. They've taken Hawking's calculations one level deeper than before, which is progress but hardly the final word. A broader moral is that different approaches to quantum gravity yield much the same answers. This suggests that beneath the details of the calculations are some common principles yet to be fully appreciated about how space and time are put together.

Timed Out?

I'd love to write a whole chapter on what the nascent quantum theories of gravity have to say about time machines, picking up the train of thought from Chapter 10. Even better, I'd love to report that physicists and venture capitalists have founded startup companies to build time machines for sale to the public. But alas, there's just not that much to say yet.

Loop gravity is mute, buckyspace and causal-set theory eliminate time machines by fiat, and string theory's findings are inconclusive. On the positive side, strings provide the two key ingredients of time machines. First, they can tear and reconnect spacetime, producing wormholes. Like black-hole formation, wormhole formation is a type of transition between phases of matter. Second, strings and branes permit negative energy, which can prevent a wormhole from collapsing in on itself. Acquiring these ingredients was once thought to be an insurmountable obstacle.

Putting the ingredients together is what now poses the problem. Gravity/gauge duality offers some clues. Time travel involves gravitational contortions whose particle counterparts violate basic tenets of quantum theory. For quantum theory to hold, gravity must somehow act to nip time machines in the bud. In one scenario, a black hole spinning faster than a certain threshold could act as a time machine, but the black hole regulates itself to prevent that. Whenever it threatens to breach the threshold, centrifugal forces cause it to bloat in size and slow itself down.

So string theory tends to support physicist Stephen Hawking's intuition that time machines foil their own formation. Nature keeps playing bait-and-switch, dangling the possibility of time machines before us and then snatching it back. Adventurers may have to content themselves with jumping into black holes.

The Least You Need to Know

- In string theory, black holes are to ordinary objects what steam is to water: another phase of matter.
- The new phase keeps a full record of everything the hole devours and ultimately releases radiation that encodes this record.
- The boundary of the black hole may be a two-way street, contrary to relativity theory.
- Loop gravity suggests that the singularity at the center of a black hole gets smeared out; otherwise, though, the details of the hole remain hazy.
- Time machines seem unlikely, but their viability is still an open question.

Chapter 20

Before the Big Bang

In This Chapter

- The start of it all?
- Stoking inflation
- Inflating without inflation
- The cyclic universe
- Shedding light on dark energy

Most cosmologists used to say that nothing came before the big bang, implying it was the dawn of time. Nowadays they're inclined to think of it merely as the sunrise on a new day. The universe might well have existed before, maybe in much the same form as it does now or maybe in some unimaginably different quantum version.

Time Before Time

My four-year-old daughter recently started asking where she was before she was born. Her question went deeper than the facts of life. What does it really mean to exist or not? How do we really grasp what it means for us not to exist, either before birth or after death? Our dinner-table conversation these days sounds like a Café Philosophique.

Science helps with these age-old questions by illuminating the twilight between existence and nonexistence, where one shades into the other. On one side, we have what is unmistakably a human being, and on the other, what is unmistakably not although all the pieces are there. In the middle, human qualities progressively emerge. We can start to comprehend nonexistence by adding or subtracting qualities one by one, such as the ability to recognize oneself in a mirror, to understand language, or to differentiate your spouse from a hat.

What applies to an individual person was writ large at the birth of our universe. Our personal origins are tightly bound up with cosmic ones. Our transition from nonexistence to existence actually began billions of years ago, when the particles and influences that make you "you" and me "me" began their tortuous journey. The qualities of the natural world progressively emerged. The origin of galaxies, particles, and distinct forces fall within the scope of current theories. Earlier milestones, such as the origin of time, demand a quantum theory of gravity, which itself is only gradually coming into being.

The term big bang, discussed in Chapter 9, properly refers to the ongoing expansion of which we are a part. Often people use the term to refer to time zero, when it all supposedly started. But the idea of a singular starting point comes from imagining rewinding the present expansion back to when all the galaxies we see would have been crunched into a single mathematical point. If there's anything cosmologists know, it's that we can't just extrapolate the present trend all the way back.

For one thing, if all those galaxies immediately started flying apart from that initial moment, there wouldn't have been any time for light, heat, or material to pass between them. The galaxies would have had no way to affect one another and come to a common temperature and density as astronomers observe them to have. Second, the density of matter in that mathematical point would have been infinite—a so-called singularity, like the one supposedly at the center of black holes. Presumably, quantum gravitational effects prevented the density from becoming truly infinite, in which case time either emerged less abruptly or extended back.

In short, something *must* have come before the big bang—that is, before the current period of expansion. What could it have been? And what came before that? Did time have an ultimate beginning or does it stretch back forever? Some scenarios inspired by string theory and other theories tackle the first question (the immediate predecessor); some tackle the second (the ultimate beginning); and some take on both.

Living With Inflation

The leading view is that a period of inflation preceded the current period of expansion. Before inflation, the precursors of galaxies were moving apart slower than light, so they had a chance to homogenize themselves. During inflation, cosmic expansion accelerated; everything in the universe came to move apart at faster than light; and the galactic precursors fell out of contact. Relativity theory permits this sort of faster-than-light motion, which arises from the expansion of space rather than motion through space. After inflation, expansion returned to its regularly scheduled program.

To push the galactic precursors apart required a form of energy whose gravity repels rather than attracts. Cosmologists generally attribute it to something called the inflaton. The inflaton is supposed to make the universe uniform, so it shouldn't single out a particular direction in space. That suggests it's what physicists call a *scalar field.* Unlike a magnetic field, which points in a certain direction (namely, toward the north magnetic pole), a scalar field is just a single number at each point in space which has no specific direction.

def•i•ni•tion

A **scalar field** is a special type of field that can be described by a single number at each point in spacetime, as opposed to both a number and a direction.

Because the Standard Model of particles has nothing that quite fits the bill and because the inflaton operated only under the extreme conditions that prevailed billions of years ago, physicists reckon that the inflaton had something to do with the unification of physics. String theory and other candidate quantum theories of gravity offer several possibilities as well as alternate ways to achieve the same goals.

Stringy Inflation

If a higher-dimension space alien tapped you on the shoulder, what would it feel like? You'd sense that something was acting on you, but you wouldn't be able to tell where it was coming from. When a force reaches us through an extra dimension, we perceive it as a scalar field, which doesn't point in any particular direction. String theory predicts many scalar fields, which represent various aspects of extra-dimensional geometry. One of them might have played the role of inflaton.

Broadly speaking, this could have happened in two ways. First, the crinkled extra-dimensional space could morph. Like everything in nature, it sought to minimize its energy. If it started off with too much energy, it reshaped itself until it reached the lowest energy it could—a "valley" in the "landscape" of possible stable shapes (see Chapter 15). As it thrashed about, its energy drove inflation. After it settled down, the universe expanded normally. Because the lowest energy is not zero energy, the universe was left with a residue of energy. This residue could be dark energy, the unidentified stuff that is now causing cosmic expansion to accelerate again.

Alternatively, space may be filled with parallel universes living on branes which scurry around exerting forces on one another. We can even get a situation where branes and antibranes, like particles and antiparticles, attract each other and, when they hit, annihilate in a blaze of strings and smaller branes. Their attractive force could have acted as the inflaton. Some of the debris of the annihilation could have wound up in our universe, thereby seeding it with matter.

Like conventional inflation, these shenanigans don't have much to say about how long the inflationary period lasted or what preceded it, except that *something* had to in order to ignite it. The starting conditions for inflation may have arisen purely by accident. In the brane scenario, for example, the branes could have played bumper cars for ages until one of those collisions triggered inflation. Once inflation began, it spread in a chain reaction.

String Gas and Black Hole Fluid

The preceding scenarios retain the basic concept of inflation and seek to describe it in greater detail. String theory also suggests more radical options that replace inflation with some other process that has much the same effect.

According to one, the universe was once filled with a gas of strings. It looked like a food fight in an Italian restaurant, with spaghetti and fettuccine flying every which way. The strings' momentum produced pressure, which tended to push outward. Because of the extreme conditions, distinctively stringy properties also came into play—namely, some of the strings wrapped around space and pulled inward. These competing tendencies balanced each other and caused the universe to remain the same size.

Because the universe didn't grow, particles weren't being pulled apart, so nothing stopped them from interacting out to large distances. They spread out into a uniform gas—or, rather, a nearly uniform gas, with the random density fluctuations we find in any gas. Eventually, the wrapped strings loosened their hold on space. Space began to unfurl; the gas cooled off; and the fluctuations seeded galaxy formation. Galactic precursors separated by more than a certain distance got caught out and became unable to interact with one another. Only later, as time passed and the expansion mellowed out, did they come back into contact.

Whenever cosmology sounds baffling, we can go back to the rubber-band model for enlightenment. Imagine space as a rubber band inhabited by two ants. In inflation, the ants begin side by side, and a vigorous pull on the rubber band carries them apart faster than they can walk. In the string gas, the rubber band is initially slack and the ants can wander back and forth at will. Then we start pulling on the rubber band. If the ants are already separated, even a fairly lackadaisical pull can carry them apart faster than they can walk.

In another scenario, the universe started off as a gurgling soup of black holes. What an unappetizing brew it was! An equilibrium developed as black holes evaporated into radiation and radiation turned back into black holes. The principle of locality broke down because it was impossible to identify specific locations in space against the featureless faces of the black holes.

But like everything, this toxic gruel had its quantum fluctuations, creating bubbles of comparatively low density. If such a bubble was lopsided, the holes inside it collided and the fluid thickened again. But if the bubble was uniform, the black holes evaporated without forming again, and a universe took root. By this reasoning, our universe is nearly uniform because otherwise it couldn't exist at all.

The Cosmic Inflection Point

Could the two big questions of cosmology, what made the universe uniform and what the big-bang singularity was, be related? One of the oldest efforts to apply string theory to cosmology, the *pre–big bang* scenario, argues just that. It takes its inspiration from the funny way distances behave on small scales (see Chapter 16). If we shrink a string, it reaches a minimum size and then acts as if it expands again. Maybe the same thing happened to the entire observable universe. Its present expansion (the big bang) may have been preceded by a period of contraction (the pre–big bang). Instead of a singularity, the universe went through a transition.

def•i•ni•tion

The **pre–big bang** scenario supposes that the universe before the big bang singularity was, in many ways, a mirror image of the post-bang one.

In many ways, the pre-bang epoch was a mirror image of the bangian one. Because the universe will continue for an eternity into the future, it began an eternity in the past, and because it will be empty in the infinite future, it was empty in the infinite past—it began a void and will end one. One thing that didn't change is that space never stopped expanding. Instead, what changed was its rate of expansion, from acceleration to deceleration. Because light couldn't keep up with the acceleration, the visible domain shrank during the pre-bang epoch.

In essence, this scenario says cosmic inflation was not a brief interlude but the entire first half of cosmic history. For cosmologists, the primordial emptiness is appealing because emptiness is the least contrived state of space. The universe was devoid not just of matter but also of force. In string theory, the strength of forces can vary, driven by the string vibrational pattern called the dilaton (see Chapter 12). In the infinite past, their strength was zero, which was an unstable condition. The forces spontaneously strengthened, as if the cosmos were switching on its own lights.

Most of the action took place just before and after the transition, when conditions were most extreme, and that's the main difficulty with this scenario. Theorists can't verify whether it really works because it's too hard to follow the universe through the transition.

Follow the Bouncing Brane

One scenario goes even further, introducing an alternative to inflation and the big bang singularity as well as relating them to dark energy. It resembles the colliding brane picture explained earlier, except that the brane doing the hitting is ours. The big bang began when our universe and another brane banged into each other like cymbals. The energy of the collision filled space with hot primordial soup, a process that the inventors of the scenario call *ekpyrosis.*

The branes rebounded but still exerted a force on each other, as if connected by a multidimensional spring. The force slowed them down, stopped them, and pulled them back together. A trillion years from now, the branes will bang into each other again, and so the cycle continues. Each go-around is much the same as the last. The universe refills with galaxies and with new forms of life that say to themselves, "All this has happened before and will happen again."

def•i•ni•tion

Ekpyrosis is the idea that the big bang began in a blaze when our universe hit a parallel one.

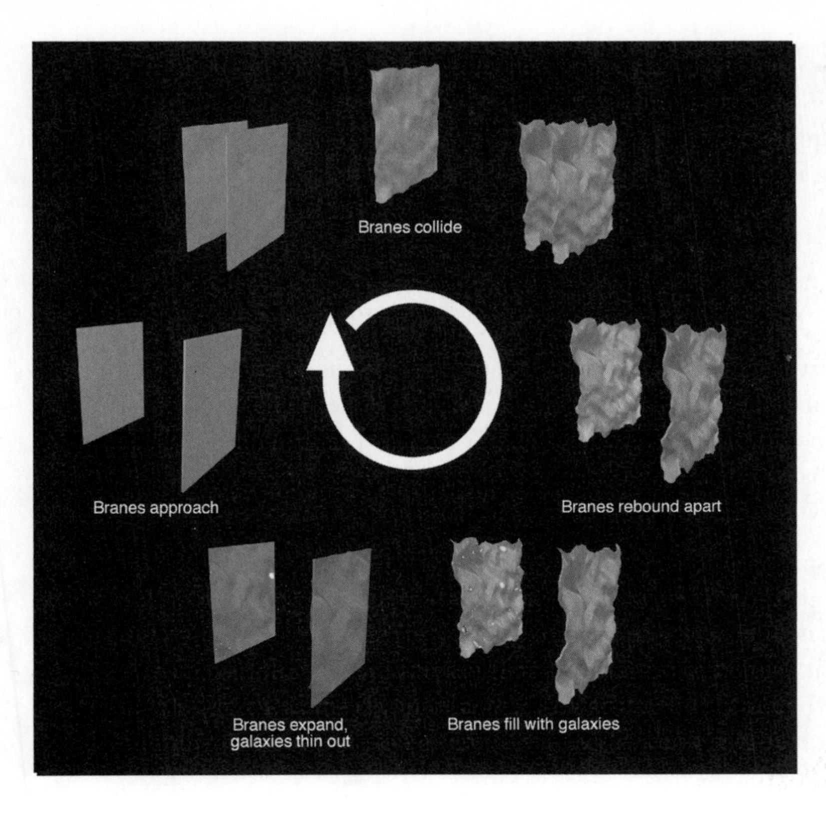

The brane containing our universe and a parallel one may collide repeatedly in an eternal (or at least long-lived) cycle.

(Courtesy of Paul Steinhardt)

The idea of a cycling universe seems to cycle around every now and then. In older versions, cosmologists supposed that the universe would expand, stop, and contract again in a big crunch. However, this concept had a fatal flaw. Whereas a big bang is nearly uniform, a big crunch is lumpy, gummed up by all the stars, planets, and other muck that formed in the interim. So the crunch wouldn't reset the universe to its initial conditions, and no true cycle could develop.

The new cyclic scenario solves that problem. The spring that connects the branes stores energy, which acts as dark energy, causing each brane to expand laterally at an accelerated rate. The accelerating expansion dilutes the muck, so that each bang begins afresh. Once acceleration ends and the branes start to approach each other, they shrink slightly and grow smoother and flatter, except for tiny wrinkles that will serve as the seeds of galaxies in the next cycle.

As in the pre–big bang model, the challenge is to follow what happens during the moment of contact. Do the extreme conditions undo each cycle's careful preparation for the next? Physicists also debate whether the cyclic model really eliminates the need for an ultimate beginning. Skeptics argue that the cycles don't precisely repeat, so they couldn't have been going on forever.

Loopy Cosmology

Cosmology in loop quantum gravity isn't as well developed as its stringy counterpart but has a certain minimalist elegance. If spacetime is a mosaic of discrete atoms, there's a limit to how much energy it can hold. If we try to pack in more, gravity turns from a force of attraction to one of repulsion, pushing the energy right back at us. This provides a natural way to put the "bang" into the big bang.

An implosion may have preceded the bang. Then as the material got ever more packed together, gravity became repulsive and turned the implosion into an explosion—it was the big bounce. In fact, the repulsive gravity could have led to a brief period of accelerated expansion as in inflation. Even without inflation, the universe had plenty of time prior to the bang in order to make itself uniform. Like the stringy scenarios, the loopy picture faces the challenge of following events through the switch-over.

Loop gravity also inspires speculation about alternatives to inflation—unconventional ways for far-flung regions to reach the same temperature and density. Maybe nonlocal links between points in space (the disordered locality mentioned in Chapter 18) serve as secret passages. Particles could sneak through them and find themselves billions of light-years from where they started. A flow of particles could have evened out the density and temperature of the early universe, despite the chasm of space separating the precursors of galaxies.

Or maybe the doubly special relativity mentioned in Chapter 16 mimicked inflation. If the Planck energy scale is a universal energy limit, the speed of light may not be constant. If it were larger in the past than it is today, light and other influences might have been fast enough to cover the distance between two distant regions. Later, as light slowed, the regions fell out of touch. After all, inflation requires galactic precursors to move faster than light, and we can achieve this either by pulling them apart or by changing the speed of light.

You can see from this discussion that theorists' imagination is expanding about as fast as the universe is. The list is encouraging because just a few years ago, the conventional picture of inflation was the only game in town. In science as in economics, competition can be a healthy thing. Within a few years, new observations should be able to start winnowing down the ideas.

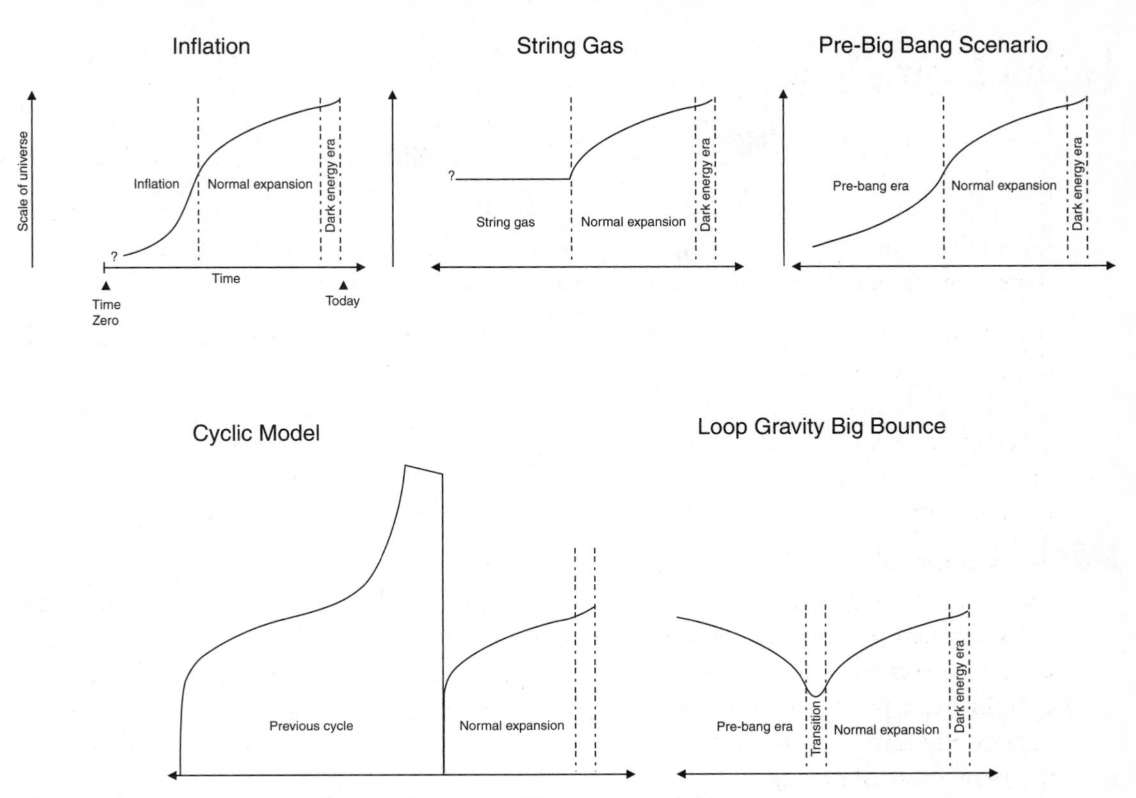

How the universe grew in a selection of scenarios (shown very roughly).

Creation *Ex Nihilo*

Whenever we talk about the origin of time, we find ourselves getting tongue-twisted talking about "emerging" when there was no time to do any "ing"-ing in. Many of the above scenarios try to get around this problem by saying that the universe has always existed. This strategy has certain conceptual advantages, such as allowing the universe to "start" in the infinite past as a placid void rather than a gnarled high-energy mess. Still, it smacks of philosophical passing-the-buck. An eternal universe requires explanation as surely as a finite one does because time must itself be explained. A theory must be able to account for the initial conditions of the universe even if they were set an eternity ago.

In the Loop

An infinite span of time has to be created in the same way as a finite span of time.
—Edward Harrison, University of Massachusetts, Amherst

Of the current approaches to a quantum theory of gravity, buckyspace comes closest to showing how time can progressively emerge. Like efforts in the 1980s by Stephen Hawking and others, buckyspace describes spacetime as a quantum average of its myriad possible shapes. At the dawn of time, we couldn't take the average but had to think in terms of all the individual shapes. Quantities such as distance and duration, which are themselves averages, did not yet have meaning.

Physicists sometimes describe this process as popping out of nothingness, but the "nothingness" they refer to was not an absolute void. All the ingredients for space and time as we know it were present; it was just a matter of putting them together. Where did the ingredients come from? What would we see if we ran the clock back to those first moments? What did it really mean for the universe to go from nonexistence to existence? No one yet has a convincing answer to a four-year-old's question.

Darkness Falls

Next to the origin of time, the puzzles of dark energy and dark matter can sound downright prosaic. But they're still among the biggest mysteries in modern science. If anything, they confront physicists more directly. The origin of time may be a far-off event, but dark matter can flutter through laboratories here on Earth, and dark energy is tearing apart our universe as we speak. Neither fits into the Standard Model of particles. These dark things are among the most visible signs of a deeper level of reality.

When dark energy burst into the world of science in 1998, string theorists seemed nonplussed. They were hardly alone; dark energy posed a challenge to all of physics. Nowadays, though, theorists can explain the current acceleration in essentially the same way as inflation: energy associated with the shape of space. It fits right into many of the scenarios I described earlier.

Or it might also arise from other spacey effects. In Chapter 14, I stated that particles transmitting the force of gravity can be trapped within our brane as long as they are neither too short nor too long. If they fall outside this range, they leak into the higher dimensions of space. The leakage puts our brane into tension, warping it. We can't see the warpage directly, but we still perceive it as a residue of energy built into space itself—namely, as dark energy.

The lesser-known approaches to quantum gravity also suggest that dark energy is tied into the basic structure of space and time. In the buckyspace theory, dark energy arises organically from the collective behavior of the primitive building blocks of space. In the causal-set theory, dark energy is a natural by-product of chopping the spacetime continuum into discrete points.

The fluids-and-solids approach to quantum gravity can explain one odd aspect of dark energy: why its density is so much lower than standard quantum theory predicts. As explained in Chapter 7, the irrepressible undulations of all the quantum fields in the world should endow spacetime with energy, acting as dark energy. These undulations are so intense that their enormous energy should rip us all to shreds. So why don't they? The fluids-and-solids theorists reason that the effect of these fields is not to endow spacetime with energy but to create spacetime to begin with. In other words, standard quantum theory double-counts the effect of these undulations. So the default amount of dark energy is zero, rather than apocalyptically huge.

Compared to dark energy, dark matter is a piece of cake. In string theory, it's one of the supersymmetric particles (see Chapter 17). To be sure, not every physicist likes cake. Some argue there is no dark matter. The anomalous celestial motions astronomers attribute to dark matter may instead reflect a breakdown of the laws of gravitation on cosmic scales. None of the current quantum theories of gravity predict such a breakdown, and the evidence seems to be against it. But if such a radical possibility were confirmed, it would transform physics and give our family something new to talk about over dinner.

The Least You Need to Know

- Both string theory and loop gravity suggest that the big-bang singularity was not the beginning of the universe but a transition from a pre-existing phase.
- The motion of branes or resculpting of space might account for cosmic inflation and dark energy.
- The present phase of cosmic expansion may be only the latest in a long-lived cycle.
- Loop gravity suggests the big bang was a big bounce.
- Dark matter might be one of the new breeds of particle predicted by supersymmetry.

Chapter 21

Ten Ways to Test String Theory

In This Chapter

- Is string theory testable?
- What does being testable even mean?
- The Large Hadron Collider
- A sampling of other instruments

Contrary to popular belief, we can test string theory and other quantum theories of gravity experimentally. Ongoing or upcoming experiments might not be able to rule any of these theories out, but negative results could cast such doubt on them that many physicists would move on. The next few years will be a time to remember in fundamental physics, as researchers push into the unexplored territory beyond the prevailing Standard Model of particles.

Testing Times

Physicists suffer the curse of their own success. General relativity and standard quantum theory do such a good job of explaining the universe that whatever underlies them must have escaped our attention so far. Either the clues are tiny, or we're so conditioned by our current worldview that we've failed to recognize something right in front of our eyes.

Many people go so far as to say that testing a quantum theory of gravity is beyond human capability—and always will be. The world's biggest particle accelerator, the Large Hadron Collider, slingshots particles to an energy of 7 trillion electron-volts (TeV). It probes features as small as 10^{-20} meter. The Planck length, where quantum gravity reaches its full strength, is a quadrillion times smaller. The grand unification scale, where all the forces except gravity merge, isn't much different. Juicing up particles to the corresponding energy would take an accelerator bigger than the solar system, the Milky Way galaxy, or the known universe (it depends on whom you ask). Congress would probably balk at funding such a thing.

But there are other ways. Broadly speaking, potential experimental tests fall into several categories:

- **Ridiculously sensitive equipment.** Some instruments might indeed be able to pick up a one-part-in-a-quadrillion discrepancy from current laws. For instance, ultraprecise laser-rangers might be able to discern deviations in the moon's orbit or quantum gravitational fluctuations in the distance between objects in a laboratory.
- **Knife-edge processes.** A housefly is way too light to register on a bathroom scale, but if it landed on your nose while you were on a perfectly balanced seesaw, it might tip the balance in your favor. Similarly, some particle processes, such as those involving so-called kaon particles, are so finely balanced that a very slight deviation can tilt them one way or the other.
- **The cosmic enlarge button.** If you set the enlargement on a photocopier to 200 percent, make a copy, make a copy of the copy, and do this 20 or so times, you might expect to make pictures of atoms. This trick doesn't work in practice (the image gets washed out), but a cosmic version just might. The expansion of the universe could have blown string-sized features up to light-years across or even larger. A black hole does something similar. Light escaping from its perimeter can get stretched from unobservably small wavelengths to a visible glow.
- **The power of multiplication.** A small effect times a very big number equals something you might see. Subtle discrepancies, allowed to accumulate over vast spans of cosmic time or distance, become measurable. A tank of water has so many particles in it that even a very rare process is a regular occurrence.

- **Moiré patterns.** Maybe we can intuit the fine-scale structure of space from the large-scale patterns it generates. Although we can't see the tiny dots of ink on this printed page, we know they're there by how they create image artifacts such as a Moiré pattern. Although we normally aren't aware of TV pixels, they become obvious when a presidential candidate wears a striped tie or paisley blouse and viewers see it as a distracting psychedelic pattern.

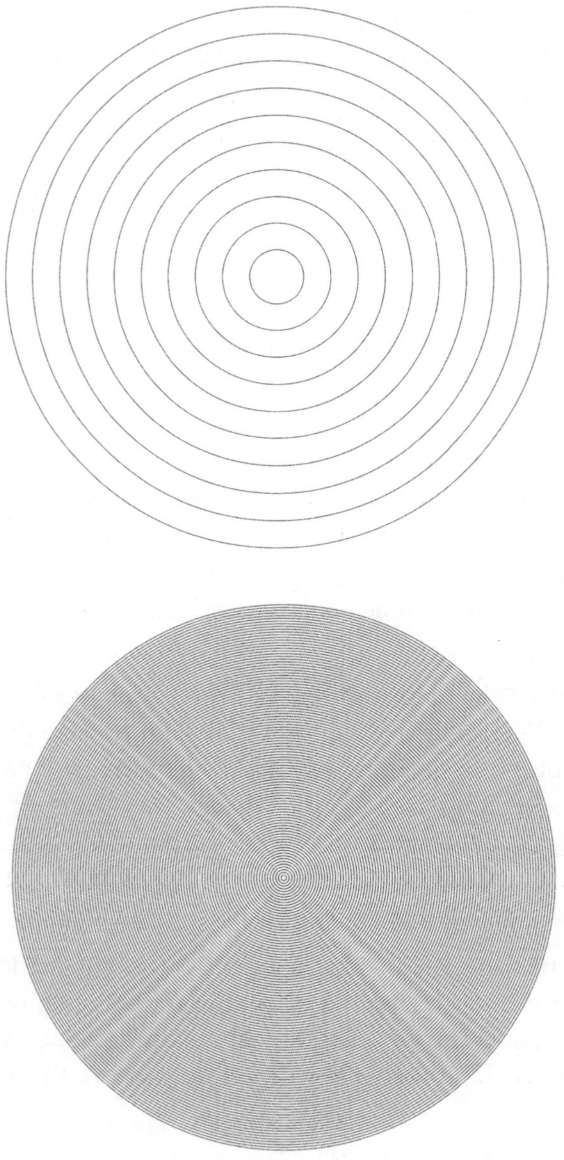

The tiny dots of ink used to print this book are too small to see, but you can detect them because of a Moiré pattern—the shadowy V-shapes on these concentric circles.

- **Hidden reservoirs of strength.** The whole business about the Planck energy presumes that gravity is as weak on fine scales as it is on large scales. Physicists need to pump up the energy of particles to compensate for gravity's innate weakness. But what if gravity naturally gained in strength on fine scales? That would happen if space had extra dimensions and if those dimensions were unexpectedly large in size (see Chapter 14). Then instruments would be able to see quantum gravitational effects directly.
- **Matters of principle.** The principles that operate under extreme conditions might carry over into the everyday world. For instance, we don't need to rocket through space at light speed to see Einstein's special theory of relativity in action. We could be sitting still and get hit by some antimatter. (That's not as bad as it sounds. Antimatter is used in medical imaging such as PET scans.) The existence of antimatter is a consequence of the principles of relativity applied to quantum particles. Likewise, a unified theory of physics might predict qualitatively new phenomena. Cosmic dark matter and dark energy could be among them.

So let's look at a sample of specific experiments that put these ideas into practice.

What Is Proof?

Notice I've been talking about "testing" unified theories, not "proving" them. Proof in science is a funny thing. When scientists talk about their ability to prove a theory, they actually care more about their ability to *dis*prove it. A theory should take a stand. It should make a prediction that can be demonstrated false. If it can weasel out of any contrary findings, what use is it?

String theory and other candidate theories make firm predictions of this sort, but unfortunately these predictions lie outside our reach—not because of any failing in the theories per se, but because quantum gravitational effects are so intrinsically weak. Even if current experiments find nary a hint of strings, they won't rule out string theory. For example, string theory predicts new particles, but if physicists don't find any, all they'll be able to say is that those particles are too heavy for current technology to create. The particles may or may not really be there.

This wiggle room troubles some scientists, especially those who are lucky to work in fields where experimental tests are cut-and-dried. But the situation is actually fairly common in science. Testing a theory is seldom like flipping a light on or off. It's more like using a dimmer switch. If a theory passes a test, we turn up the dimmer, and if it fails one, we turn it down. When the room gets too dark, scientists drift off to a brighter one.

So although strictly disproving string theory is beyond our capability right now, a series of successes over the coming years would encourage string theorists that they're on the right track. An accumulation of disappointments would persuade them to look for other options.

1. The Large Hadron Collider

The world of particle physics right now revolves around the Large Hadron Collider (LHC), the world's most powerful and most expensive hammer. It smashes protons into little bits to see what they're made of and to give new types of particles a chance to form. The collider flings particles to within 10 kilometers per hour of the speed of light, for a kinetic energy of 7 TeV. By comparison, the previous record-holder, the Tevatron, achieved 1 TeV. The LHC pushes into an energy range where the Standard Model gets ratty. Most physicists think a new theory has to show up. What will it be?

How the Collider Works

The machine starts by breaking protons out of gaseous hydrogen atoms. Because protons are positively charged, they flock toward anything with a negative charge. By setting up a series of electric fields, physicists catapult the protons to near light-speed in stages. Four accelerators of progressively increasing size boost the particles, which then enter the main accelerator ring like cars merging onto the Washington Beltway.

This ring is about 8.5 kilometers across and consists of an underground tunnel the width of an airport jetway running under farms on the outskirts of Geneva. Two pipes carry beams of protons circulating in opposite directions. Magnets around the pipes steer the protons in a circle, with the protons making 11,000 loops a second. Working in a different mode, the LHC can also smash together atomic nuclei.

Protons collide head-on at one of four points along the ring, each a giant, heavily instrumented cavern. Some 50 million collisions occur per second, each producing thousands of debris particles that spray out in all directions. Particle detectors nested in concentric layers register their direction and energy. They collect a DVD's worth of data every five seconds—such a flood of data that physicists and engineers have had to develop new parallel-computing and networking technology to process it.

Getting around the 27-kilometer-long Large Hadron Collider presents its own challenges; people who enter the tunnel also have to carry emergency breathing apparatus.

(Copyright 2005 CERN)

The LHC represents the pinnacle of human technology. Some 6,000 scientists work on it. In all, it cost 10 billion Swiss francs (about $8 billion). Of this, the United States kicked in about $500 million. That's a lot of money, but in terms of the energy delivered to a proton, the collider is extremely cost-effective. If an ordinary hammer were as efficient, it'd cost a millionth of a cent.

I visited the LHC when its instruments were about halfway built, so that I could get a sense of their true scale before they completely filled their caverns. One thing that struck me was the combination of finesse and power. Imagine the intricacy of a Swiss watch filling a space the size of a heavy industrial factory, crisscrossed with cranes, girders, and gangways. Just keeping track of all the wires and cables is monumental.

What It Looks For

The goal of the LHC is to create particles that human beings have never seen and whose existence reveals a new layer of physical reality:

- **Higgs particle.** The marquee attraction is the Higgs particle or something else that serves the same purpose—namely, to explain why the weak nuclear force is so weird (see Chapter 5). The Higgs is the archetype of a new type of substance that differs from ordinary matter and force fields—namely, a so-called scalar field. Other such fields may have played a decisive role in making the universe what it is today.

- **Sparticles.** A key prediction of string theory (though not only of string theory) is supersymmetry, according to which every particle has a partner particle or sparticle (see Chapter 17). If the collider doesn't find sparticles, it technically won't rule out string theory but will make people wonder. Conversely, the discovery of supersymmetry won't disprove alternatives such as loop gravity but will make them that much less appealing.
- **"Missing" energy.** The lightest sparticle might account for the universe's dark matter. Astronomers don't know a great deal about dark matter, but one thing they do know is that it passes right through ordinary matter as if it weren't even there. So if the LHC manages to create some dark particles, they'll slip through the walls of the collider and escape (harmlessly) into the cosmos. The particle detectors won't register them directly but will notice that some energy had been lost.
- **Extra dimensions.** If space has extra dimensions and if those dimensions are large enough, the LHC can detect them. Ordinary particles moving through space could rattle in the extra dimensions like a car driving over a crack in the road. We'd perceive this rattling as a new set of particles. Some particles might escape into the extra dimensions altogether, which the particle detectors would register as missing energy. Most dramatically, the force of gravity could strengthen at the energy levels achieved by the LHC. Energetic particles might collapse into tiny, short-lived black holes or even wormholes. These exotic objects would quickly disintegrate in a distinctive burst of particles.

All Tangled Up

Would creating black holes in the lab be dangerous? Normally we think of black holes as ravenous monsters, but the small ones are pitiable, tortured souls. By their very nature, they're unstable. They go pop before they can do any damage. They'd pose a threat only if they had an exotic kind of "charge" in addition to electric charge and quark color. The force exerted by this charge would counterbalance gravity and stabilize the hole. But the protons that create the hole don't have this charge, so the hole can't either. Even in an absolute worst-case scenario where theorists were wrong and a hole started to devour matter, it would take billions of years to grow to dangerous proportions.

- **The unknown.** The only thing that makes physicists happier than confirming a prediction they made is to find something they never anticipated. Every time they've cranked up the energy of their instruments, they've found something unimagined: new particles, new forces, new processes. Whatever it is, it'll transform humanity's understanding of the universe.

Quantum Leap

Particle physics projects take years to plan, so physicists are already looking beyond the Large Hadron Collider to an even more powerful machine, the International Linear Collider. It would smash electrons rather than protons. Electrons, being simpler particles, produce tidier collisions and more definitive data. The hang-up is that they're lighter in weight, so they tend to lose whatever energy we give them. For this reason, the new collider would be linear: particles moving in a straight line are less prone to lose energy than those moving around a circle.

2. Testing Dr. Einstein

On close examination, the continuum of space could dissolve into a foam (see Chapter 16). Long-wavelength light beams are oblivious to this fine filigree, but some theorists think that a short-wavelength beam would bounce around inside the foam and take either longer or shorter to make its way through space. If so, special relativity—which assumes that all light beams travel through empty space at the same rate—may break down on microscopic scales. Most versions of string theory assume that special relativity holds all the way down, so a violation would call string theory into question.

Gamma rays are the shortest-wavelength type of light. Astronomers routinely study gammas from black holes and exploding stars billions of light-years away. These sources tend to flicker. Each flicker releases gammas of various wavelengths, which then race across the universe for billions of years—plenty of time for a minuscule speed difference to add up. By the time they reach our telescopes, the gammas could be substantially out of sync.

In 2007, a group of physicists using a ground-based gamma-ray telescope named MAGIC announced just such an effect. They found that shorter-wavelength gammas from a black hole arrived about four minutes before longer-wavelength ones. This is one of those "extraordinary claims require extraordinary evidence" situations. It's hard to distinguish a genuine speed discrepancy from more prosaic effects, such as flickering that isn't exactly synchronized across wavelengths. It will take more than one black hole to tell. NASA's new Gamma-Ray Large Area Space Telescope (GLAST) satellite will analyze the light of multiple celestial bodies for this effect.

3. Catching Some (Cosmic) Rays

Nature can fling particles to speeds that accelerator-builders only dream of. Our planet is continually bombarded by particles (probably mostly protons) known as *cosmic rays*. The highest-energy ones pack a punch 10 million times greater than the Large Hadron Collider can manage. They move so close to the speed of light that the difference is out in the twenty-second decimal place.

Earth's atmosphere thankfully protects us from these particles. When a cosmic ray hits the top of the atmosphere, it shatters into less-energetic particles, which in turn break up in even less-energetic ones, eventually showering the ground with billions of electrons, gamma rays, and other particles. We get a mild radiation dose from these and lesser-energy cosmic rays but nothing like what astronauts on deep space missions would have to deal with.

def•i•ni•tion

Cosmic rays are particles such as protons zipping through outer space at high speed.

The world's leading cosmic-ray observatory, the Auger Observatory in the ranchland of western Argentina, detects the debris particles in two ways: as they give off flashes of light in the air on the way down and as they pass through a network of detectors on the ground. By piecing together the debris like airplane-crash investigators, physicists reconstruct what the original particle was and where it came from.

Physics-loving cattle gather around one of the Auger Observatory's 1,600 particle detectors, a 3,000-gallon water tank.

(Courtesy of Pierre Auger Observatory)

Though potent, the highest-energy cosmic rays are fairly rare. Auger and other observatories are lucky to see a handful of the most energetic ones per year. Their main significance is their mere existence. What creates them? Are slow particles slingshot by shock waves or strong magnetic fields, perhaps driven by giant black holes? Or are the particles spit out by exotic objects, such as those predicted by string theory?

Moreover, how do these powerhouses make their way across space? Special relativity predicts that to such a fast-moving particle, the tenuous radiation that fills space might as well be a thick sea. A powerful particle rapidly exhausts itself trying to wade through it. If the sources of these particles are too far from Earth, something must be wrong with special relativity or else the particles wouldn't have made it here.

Over the past several years, various observatories have sought to trace the particles' paths back to their sources. In 2007, however, Auger settled the issue by matching up energetic cosmic rays with galaxies that are fairly nearby. So it looks like special relativity holds after all.

4. Written on the Sky

The farthest back in history we can see is the microwave background radiation, released at a cosmic age of about 400,000 years. Before then, the primordial soup was opaque to light. But it wasn't opaque to gravitational waves (see Chapter 11). If astronomers could somehow observe these waves, they'd be able to look still further back into cosmic history. They might peer all the way back to a time when the forces of nature were unified and maybe even to a time before the big bang began.

The best hope right now for seeing these ancient gravitational waves is indirect: by the way they affected the microwave background. Waves rippling through the primordial soup kneaded it, alternately increasing and decreasing its density. The microwave background provides a snapshot of where things stood at 400,000 years. In the scheme of things, gravitational waves were a fairly minor player, but they had a corkscrewlike motion that gave the matter they passed through a telltale swirl. Nothing else did that.

The swirl, in turn, twisted the microwaves of the background radiation. It altered the direction that the microwaves oscillate in—their so-called polarization. This polarization signal is extremely weak, but the European Space Agency's new Planck Surveyor satellite should just about be able to detect it. Ground-based and balloon-based telescopes will follow up with even greater sensitivity.

One potential source of gravitational waves was the process of cosmic inflation. If inflation occurred in the period of grand unification, it generated fairly strong waves.

Alternatively, inflation may have to do with string theory (see Chapter 20). Stringy processes generate hardly any gravitational waves. So the presence or absence of polarization will help pin down whether inflation occurred and what caused it.

The microwave background is a good place to go hunting for other clues, too. The background is polka-dotted with spots, which are the precursors of galaxies and larger celestial bodies. Cosmologists think the spots used to be teeny-tiny quantum energy fluctuations that cosmic expansion blew up to enormous size. The effects of strings may likewise have been enlarged, and if so, they'd alter the pattern of spots.

The microwave background might even reveal one of the most profound aspects of quantum gravity, the holographic principle (see Chapter 18). The principle says that the amount of information in a system is limited. The entire observable universe counts as a system. Early on, it may have passed through an information bottleneck, limiting its information content to about a gigabyte. High-resolution images of the microwave background might find that it's literally broken into pixels like a giant computer screen.

5. Gravitational Wave Detectors

Besides looking for gravitational waves indirectly in the cosmic microwave background, astronomers have also sought to observe them directly. In recent years, a number of gravitational-wave detectors have sprung up in the United States, Europe, and Japan. The most sensitive is the Laser Interferometer Gravitational Wave Observatory, which has twin stations in eastern Washington state and southern Louisiana.

Each observatory consists of two arms in an L-shape. An ultraprecise laser-ranger monitors the length of each arm, looking for the distinctive squeezing and stretching that a passing wave would induce. A ground-based observatory tends to pick up short-wavelength gravitational waves, and its main objects of study are dense stars and star-sized black holes.

If the universe existed before the big bang, it might have seeded space with short gravitational waves. The observatories aren't yet sensitive enough to see them, but upgraded versions might be. Then humans could see deep into our cosmic prehistory.

The observatories might probe quantum gravity in a completely different way, too. Their laser-rangers are so precise that some physicists think they could detect oscillations in distance caused by quantum gravitational fluctuations. In other words, they could see the spacetime foam directly.

6. Watching Protons Fall Apart

According to the Standard Model, protons are immortal. We can destroy one by smashing it, but it won't die of its own accord. The reason is that a proton is already the lightest possible combination of three quarks, so swapping in a different type of quark only increases its energy. Some outside force would have to make up the energy difference, so the process can't happen spontaneously. The only way out would be for one of the quarks to transform into an entirely different type of particle, but the Standard Model doesn't allow that. Once a quark, always a quark.

A grand unified theory changes matters. It breaks down the distinctions between quarks and the other main type of matter particle, leptons (see Chapter 7). So a quark can, and eventually will, transform into a lepton. When this happens to one of the quarks inside a proton, the proton falls apart.

Current versions of these theories, based on string theory, predict that protons last 10^{35} or so years. That doesn't mean that protons have little alarm clocks in them that go off after 10^{35} years, whereupon the particles all give up the ghost. Rather, there's a chance that any proton will decay at any moment—but only a very small chance. If we'd watch 10^{35} protons for one year, the theories predict we'd see one of them kick the bucket.

The current record-holder is the Super-Kamiokande detector in central Japan. It's an underground water tank over 10 stories tall, holding 50,000 tons of water. Light detectors ring the tank, watching for the telltale flash of light that signals a proton's demise. Their failure to see any protons fall apart means the particles must last at least 10^{33} years.

To look for much rarer decays will require a tank of water 10 times bigger than Super-Kamiokande, and teams of physicists in Japan, Europe, and the United States have proposed several possible designs. If physicists ever do see a proton die, it will be the rarest event that humans have ever witnessed.

7. Seeing Dark Matter

Even if the LHC manages to create dark matter particles, physicists won't be able to study them in detail because the particles will immediately fly out of the accelerator. To deduce the properties of the particles, physicists will need to compare notes with astronomers. Unfortunately, astronomers don't have a lot to say right now. All they know is that visible celestial bodies aren't moving as expected, indicating that some unseen material is yanking on them gravitationally.

If scientists look closely enough, dark matter might reveal itself in ways other than gravity. As the solar system moves through our galaxy, it should encounter a headwind of dark matter. Huge numbers of dark particles should stream through our planet. Particle detectors on Earth might be able to pick them up. Every now and then, one of the particles could score a direct hit on an atom in a detector, knocking off some electrons and heating up the detector ever so slightly.

The stumbling block is that particles from other sources, such as naturally occurring radioactive materials, have much the same effects. Detector-builders are trying various tricks to pick out the dark particles. For example, they combine detectors of multiple types, which respond to different particles in distinctive ways.

Another way to see dark matter is to scan the skies for collisions between dark matter and dark antimatter. As always with matter and antimatter, the two annihilate each other and give off a pulse of gamma rays, which gamma-ray observatories such as GLAST might pick up.

Identifying the composition of dark matter is just the start. Being able to see dark matter would open up a whole side of our universe that has been entirely hidden from us. Astronomers estimate there's four times more mass in dark matter than in visible matter. Who knows what worlds might lurk out there, made up of dark rather than visible stuff?

8. Cosmic Strings

The phrase "cosmic string" is almost designed to confuse. The original concept had nothing whatsoever to do with string theory. Early on in cosmic history, as the expanding universe cooled, quantum fields crystallized. They did so unevenly in space, leaving creases in them, like those annoying kinks you get when you try to wrap a plate of leftover food with clingy plastic. Physicists call these creases cosmic strings. They appear as long, skinny objects of obscenely high density. Science-fiction writers like them as a potential sources of energy for advanced civilizations.

Now it turns out that these strings might have something to do with string-theory strings after all. Pumping energy into a fundamental string could inflate it to astronomical size. One-dimensional branes also can act as giant strings. Strings of all these types would betray themselves by the way their highly concentrated gravity bends the light of celestial bodies they pass in front of. Cosmic strings might also generate gravitational waves, magnetic fields, and cosmic rays. We have seen hints of them but no confirmed sightings yet.

9. Tabletop Gravity

The LHC isn't the only instrument that might peer into extra-dimensional space. If the hidden space is sufficiently large (a big if), gravity would strengthen on microscopic scales. As the weakest of the known forces of nature, gravity is the hardest to measure precisely. It's easy enough to study over large distances, where huge masses compensate for the force's intrinsic weakness, but gets tricky over short distances.

Physicist Eric Adelberger and his team at the University of Washington in Seattle have verified that the force of gravity obeys the usual laws down to a distance of about 0.05 millimeter, about the width of a human hair. The centerpiece of their setup is a weight hanging from a thin string. It dangles over a metal disk, which exerts a slight gravitational pull on it and twists the string ever so slightly. Lasers shining off the weight can detect the twist and infer the strength of the force.

The apparatus is so precise that it can be confounded by effects as small as water accumulating on a nearby hill. It tests not only the law of gravity but also other features of Einstein's general theory of relativity, such as the equivalence principle (see Chapter 3). So far, it has found no discrepancies. Several labs are working on even more precise devices.

Another way to look for strange features of gravity is to track Earth's moon in its orbit. Physicists using the Apache Point Observatory in New Mexico bounce lasers off mirrors left by Apollo astronauts and track the moon, which is about 380,000 kilometers away, with a precision of one millimeter—a few parts in a trillion. So far, they've seen no deviations from the predictions of Einstein's general theory of relativity. The experimental precision is getting to the point where it might detect the effects of dark energy.

10. Hints of Other Universes

Almost by definition, other universes are unobservable. If we could see them, they wouldn't be other universes but part of ours. But being told that something is unobservable only makes physicists more determined to observe it. They've come up with several ways of solidifying the argument for the existence of a multiverse.

The main argument for parallel universes is the anthropic principle: the striking coincidence between many of the properties of the observed universe and the properties required for humans to exist (see Chapter 15). A coincidence begs for an explanation, and the multiverse offers one. There are a range of universes with different properties, and we live in one with the right properties for us because otherwise we wouldn't be around at all.

One way to test this idea is to see whether the coincidence is all it's cracked up to be. The most striking coincidence is the amount of dark energy. Had there been much more of it, our universe would have blown apart before galaxies, planets, and life could have formed. Making observations of the midget galaxies that surround our Milky Way galaxy could verify this claim. These galaxies formed within a few hundred million years of the start of the big bang. Even if the value of dark energy were 1,000 times greater than it is, they'd still have formed.

Do these small-fry galaxies still have planets in them? The next generation of orbiting and ground-based telescopes will be able to check. If the galaxies lack planets, that would lend support to the claim that galaxies of the size of our Milky Way had to form to allow for planets—and therefore life. But if the galaxies have planets, then the coincidence won't seem all that coincidental.

Another test is to see whether the so-called constants of nature are indeed constant. In string theory, quantities such as the strength of forces can vary, producing parallel universes with different laws of physics. Astronomers studying ancient gas clouds have indeed found hints that electromagnetic forces were weaker back then, but this work remains controversial.

The Least You Need to Know

- However compelling a theory is, it must pass experimental tests to become accepted.
- String theory and other quantum theories of gravity are now at the point where they can be tested, at least to a limited degree.
- The most crucial tests in the coming years will come from the world's biggest particle accelerator, the Large Hadron Collider.
- Other particle detectors, astronomical observatories, and orbital facilities will also test aspects of the theories.

Chapter 22

The String Wars

In This Chapter

- The vociferous critics of string theory
- A long history of discontent
- Making sense of the criticism

String theory has aroused strong feelings since it first became a viable approach to the unification of physics in the early 1980s. Some consider it the culmination of science; others, its debasement. The critics have been especially vociferous lately, taking their case to the public in lectures, books, and blogs. To what end?

String Theory and Its Discontents

Particle physicists don't call themselves high-energy physicists for nothing. The term technically refers to the high energies of the particles, but it could equally well refer to the people who study them. Compared to other physical scientists, such as astronomers and geologists, particle physicists have always struck me as people with strong personalities, strong opinions, and strong language. The atmosphere in university physics departments is more bracing than embracing. Every rule has its exceptions, of course. But by and large, they're not what I'd call laid-back types.

This culture has helped to fuel the so-called String Wars, the sometimes rancorous public debate over string theory. The discord has an edge that you seldom find elsewhere in the physical sciences. Both proponents and critics of string theory have been prone to exaggerate. If you read critics' books, you're left thinking that the best minds of our generation have been destroyed by the madness of string theory. If you read string theorists' works, you wonder why anyone would do anything else with their lives but get wrapped up in strings.

Each side feels it's the aggrieved party. Critics see string theory as a steamroller crushing all other forms of theoretical physics in its path. String theorists think their critics are the steamrollers: prominent scientists who take their case to the public rather than hash it out in scientific journals and conferences. Picking a way through this minefield isn't easy.

The 20 Years' Wars

Headlines such as "unraveling," "about to snap," and "epitaph" make it sound like criticism of string theory is a new trend. Actually, it's nothing new at all. The tone was set in the mid-1980s during the first string revolution, when string theorists' sweeping claims raised other physicists' hackles:

- In an opinion column in 1986, Paul Ginsparg and Sheldon Glashow wrote that their field threatened to break free of its experimental moorings and descend into murky contemplation "to be conducted at schools of divinity by future equivalents of medieval theologians." They added, "For the first time since the Dark Ages, we can see how our noble search may end, with faith replacing science once again."
- The renowned physicist Richard Feynman said in a 1987 BBC interview, "I think all this superstring stuff is crazy." He elaborated, "I don't like that they don't check their ideas. I don't like that for anything that disagrees with experiment, they cook up an explanation."

- In a 1990 essay, Asim Barut wrote, "These objects [strings] are so far removed from our experience that I doubt whether such an endeavor is properly part of science, or scientific method. … Science was supposed to free us from the tyranny of belief. The new trend may take us a full circle back to the domain of beliefs."

In the mid-'90s, books by two science writers highlighted these concerns: David Lindley's tightly argued *The End of Physics* and John Horgan's more polemical *The End of Science.* Both argued that physicists might never be able to prove string theory right or wrong. They'd get lost in inconclusive debates; students would choose other careers; and public interest would drain away. Fundamental physics would end not with a bang but with a whimper.

In 1999, string theorist Brian Greene published his bestselling book on string theory, *The Elegant Universe*, which led to a *Nova* mini-series in 2003. Not long after, mathematician Peter Woit, who is a particle physicist by training and temperament, started his blog, "Not Even Wrong." The name alludes to a famous put-down that physicist Wolfgang Pauli once applied to a vacuous theory (long before string theory came along). Woit has quickly become the most persistent and forceful critic of string theory.

The blog developed into a book by the same name, which came out in 2006, around the same time as another critical book by physicist Lee Smolin, *The Trouble with Physics.* This one-two punch got a huge amount of attention, both inside and outside physics. Other critics have weighed in with their own books, lecture appearances, and colorful put-downs: Lawrence Krauss, "hallucination," Phil Anderson, "futile," Robert Laughlin, "too much lipstick," and Joäo Magueijo, "masturbation."

But string theorists have not taken this criticism lying down. They have books and blogs of their own, although few respond with the same venom. The main exception is Luboš Motl's blog, which truly fights matter with antimatter. As long as you don't take his and Woit's over-the-top comments too seriously, their blogs play a very useful role in defining the extremes of scientific opinion on string theory. If you think you understand something, test yourself by visiting these blogs and looking up what they have to say about the topic. If you can understand their comments and convince yourself you know how to respond, you've passed the Woit/Motl test.

Making Sense of the Complaints

This barroom brawl shouldn't be mistaken for a mass mood swing among physicists. Physics does go through swings, but this supposed backlash isn't one of them. Rather, it's largely a public-relations affair. None of the criticism is new, and researchers have already made their choices about whether to pursue the theory or not. Few critics have actually done any work on string theory. That doesn't invalidate their criticism, but it does mean that the books and blogs aren't signs of an exodus from string theory. It just means that the critics happen to be talking louder right now. String theory is as intellectually vibrant, or not, as it has ever been.

So what are we to make of the animus? Let me paraphrase the most common arguments and offer some thoughts.

"It's Taking Too Long"

Woit writes that string theory has failed "despite more than twenty years of effort," "more than twenty years later," "more than twenty years of intensive research"—I lost track of how many times his book mentions twenty years.

These remarks remind me of what happens when I get up in the morning. I set the alarm clock saying to myself, "It can't possibly take more than a few minutes to jump in the shower, throw on some clothes, eat breakfast, read the paper, check my e-mail, play with my daughter, and catch the train, so I might as well sleep in." And every morning I rush out the door with half a bowl of cereal and an unopened newspaper left on the kitchen table. Things take time. We humans are notoriously bad at estimating how long they'll take.

There's no 20-year shot-clock in science. The United States launched the War on Cancer in 1971 and HIV vaccine development in 1984. Each time, people predicted victory within a few years. In physics, over 70 years passed between the first shot in the quantum revolution and the triumph of the Standard Model. It's a good thing Woit doesn't apply his 20-year rule consistently, or else he'd brand most of modern science, not just string theory, as an intellectual failure.

Science is hard. Why should researchers abandon a strategy they consider promising just because they can't meet a silly deadline? It's not as if string theorists have been doodling and daydreaming for those 20 years. They've been exploring a very intricate theory. It might lead nowhere, but how can they know without checking it out? Even if it dead-ends, it won't have been a waste. String theory has already cross-fertilized other areas in physics and pure mathematics. When you're talking about cutting-edge science, you need to take the long view.

"It Can't Be Tested"

At the heart of the complaint that string theory is somehow "not science" is that it has yet to make a definitive experimental prediction—numbers that one can read off a dial to prove or disprove the theory. And the critics have a point. The remoteness of theory from experiment has always troubled string theorists. And loop theorists. And buckyspace theorists. And causal-set theorists. All the proposed quantum theories of gravity are in the doghouse on this one. Gravity is weak, so its quantum features are very, *very* subtle. Even the narrower goal of grand unification, which leaves gravity to the side and tries to merge just the three forces of the Standard Model, is similarly remote. These are facts of life, rather than reflections on string theory per se.

I think we hear this complaint more about string theory than about, say, loop gravity for those cultural reasons mentioned at the beginning of the chapter. String theory grew out of particle physics, and loop gravity grew out of relativity. Particle theorists have traditionally worked closely with experimenters, whereas relativity theorists have gone decades without data. Most of the early criticism of string theory came from the old guard of particle physics, who held it to the same standard as other particle theories. They expressed their dislike with characteristic particle-physics directness. So the String Wars began as a family quarrel.

Some string theorists thumbed their nose at the old guard by describing their theory as "postmodern" physics. They meant that, lacking data, they'd have to evaluate it on other grounds such as internal consistency. But they never thought they could get away without ever having to prove it; they just reckoned it'd take time to figure out how to do that.

As discussed in Chapter 21, string and loop theorists have come up with lots of potential tests. Negative results wouldn't strictly disprove either theory, but that's par for the course. The history of science shows that new theories can seldom be ruled out so cleanly. They are gradually boxed in. Eventually they either vindicate themselves or reach a tipping point where physicists flock to the alternatives.

"It Can't Explain Anything"

String theory's original ambition was to explain the observed world as the unique outcome of a few basic principles. It hasn't quite worked out. Theorists now think the laws governing our observable universe are just our local bylaws. Other regions of space can have different bylaws. String theory gives regions wide latitude to settle into their own local arrangements. The full set of permissible bylaws is known as the landscape (see Chapter 15). Some aspects of our bylaws may have no real explanation other than they have to allow for us humans to exist or else we wouldn't be here to talk about it.

If that's the case, then what good is string theory? To critics, it sounds like anything goes—the theory makes no firm predictions, so it can never really be tested and can't be called scientific.

Yet string theory doesn't allow for absolutely any set of bylaws. It still establishes some core principles that all regions of space should abide by. If experimenters find aspects of our universe that violate those principles, they'll disprove the theory. In addition, some aspects of our bylaws might go beyond the strict requirements of our existence. They could reflect natural processes that single out certain bylaws over others. If so, string theory could at least partly live up to its original dream.

"It Presumes the Shape of Spacetime"

String theory doesn't capture spacetime in all its fluidity. To describe the force of gravity, string theorists first assume that spacetime has a certain shape and then imagine graviton particles flying around like baseballs during pregame warm-up. A full-fledged quantum theory of gravity shouldn't need to assume anything. The shape of spacetime should arise organically.

For loop theorists such as Smolin, this is string theory's main defect. It's a fair point. In the jargon, string theory isn't background-independent (see Chapter 11). On the other hand, this deficiency doesn't prevent string theorists from describing any gravitational situation by taking a series of snapshots of spacetime. It may not be as impressive as a full movie, but the difference is one of presentation. Both a movie and still frames represent the same scene. In other words, the lack of fluidity doesn't seem to be a failing of the theory per se, but of the mathematical tools that theorists use to express it.

"It Suffers from Groupthink"

Why, the critics ask, if string theory is so flawed, do so many physicists persist with it? Their answer is some personal defect of the string theorists themselves: arrogance, narrow-mindedness, or herd mentality. In defense of this claim, the critics trot out anecdotes, such as silly things string theorists have said over the years. Without a detailed sociological analysis, it's hard to know what to make of this. It's the kind of claim that people always make when they dislike what other people are doing. Indeed, string theorists have said much the same of the critics themselves.

So where does that leave us? Many people on both sides seem awfully convinced that they're right and the other side is wrong. How can they be so sure? This is quantum physics, after all—which doesn't have a reputation as the world's easiest subject.

A more straightforward explanation is that researchers have a legitimate difference of opinion. The science is complex and unsettled. There are good reasons for smart people to pursue string theory and good reasons for equally smart people to question it. Everyone makes his or her own judgment. Right now, the majority have thrown in their lot with string theory. Public hectoring won't change that. What would is if someone came along with a more compelling alternative.

"My Theory Is Better"

It's easy to criticize, harder to offer a better idea. To the extent that critics have done that at all, most advocate exploring the nooks and crannies of current theories, waiting for new data, and inching forward. And there's something to be said for that. Every now and then, a theorist discovers some unexpected phenomenon lurking within current theories—a reminder that they're incompletely understood. Such incremental efforts used to be more popular, but the cancellation of the Superconducting Super Collider pulled the rug out from under them. Waiting for new data came to look like *Waiting for Godot.* In the past couple of years, as the Large Hadron Collider has geared up, many physicists have gone back to pushing the Standard Model one step, and only one step, further.

As a way to unify physics, though, the incremental approach has come up empty-handed. The failure of earlier incremental approaches is what led so many physicists to pursue string theory. By the early 1980s, experiments had ruled out grand unified theories proposed by Glashow and others, and physicists were casting around for a more radical approach. Because general relativity and quantum theory have complementary problems, it might actually be easier to solve them by lunging for a final theory rather than trying to inch forward.

In retrospect, it turns out that even the incremental study of standard quantum theory would probably have led physicists to something like string theory anyway. According to the so-called gravity/gauge duality (see Chapter 18), souping up the Standard Model naturally gives us string theory.

Quantum Leap

If there's one thing everyone can agree on, it's that the system for funding science has become a nightmare. One way scientists get money to do their work is by applying for grants from government agencies such as the U.S. National Science Foundation. But the system is horrendously inefficient. Some scientists find they spend more time writing grant applications than doing research. Out-of-the-box ideas and young scientists are often left out in the cold. A National Academy of Sciences panel in 2005 argued for special grants to help young researchers start their careers and to enable "high-risk, high-payoff" research.

Some critics, notably Smolin, Roger Penrose, and Gerard 't Hooft of Utrecht University, fall into a different category. Their real beef isn't with string theory but with what they see as the relative neglect of other approaches. Smolin, for one, has predicted that loop gravity and string theory will eventually merge to take advantage of their offsetting strengths and weaknesses. He was a founding faculty member of the Perimeter Institute, the only place in the world where proponents of all the theories work under a single roof. For their part, string theorists are less likely than they once were to dismiss loop gravity out of hand. Many now have at least a passing familiarity with it. Greene, notably, commends Smolin's vision of an eventual merger.

Not much else has come of these olive branches, though. String theorists find loop gravity as inadequate as loop theorists find string theory. The two groups start from such different premises and come from such different cultures that they may never have a true meeting of the minds.

Most physicists are very open about the deficiencies of the theory they work on. It's a matter of emphasis. Are these deficiencies teething pains or death blows? Reasonable people can disagree. I personally think it's great that physicists take a multitude of approaches and occasionally borrow ideas from one another. My main concern is that the harshness of the debate backs people up against the wall, making them less willing to listen to constructive criticism or trawl for ideas outside their own subfield. As a result, the world might have to wait that much longer for a quantum theory of gravity to emerge.

The Least You Need to Know

- String theory has always had prominent critics, although they've gotten more vocal lately.
- Critics say the theory is taking too long to develop and doesn't make firm experimental predictions.
- Many of the complaints have an element of truth, but apply not only to string theory but also to all the proposed theories.
- One thing is undoubtedly true: all the proposed theories are still works in progress.

Chapter 23

What Now?

In This Chapter

- What's missing?
- The nature of time
- The status of quantum theory
- The eternal mystery

String theory, loop gravity, and the other theories I've talked about are all still works in progress. Even their staunchest proponents think they're missing something. But what? And how will physicists suss it out? Idea-swapping, sophisticated experiments, and philosophical reflection will all lend a hand.

Something's Missing

Our fingers can pick up things ranging from a grain of sand to a heavy suitcase. Our eyes can see objects ranging in color from red to violet. Our ears can hear sounds ranging in pitch from the buzz of a fluorescent light to the whine of a mosquito and in volume from the heartbeat of a squirrel to the exquisite musicality of a Ramones concert. But the world goes far beyond what we can directly perceive.

To paraphrase Victor Hugo, science is the microscope of thought. It lets us see what we couldn't otherwise. Through it, we see that a chair is mostly empty space and that our bodies glow with the brilliance of a 100-watt light bulb. Each level of explanation turns up the magnification a bit more. A quantum theory of gravity goes all the way. Through it, we might see space bubbling like beer head and other universes lying parallel to our own.

Such a theory might even be the "theory of everything," as is the ambition of string theory, though not so much of the other approaches. The phrase is somewhat deceptive. A theory of everything wouldn't be the theory of absolutely everything. It wouldn't answer whether we are alone in the universe or why people seem so oblivious to their own incompetence. But it would make sense of other strange aspects of the world, such as why matter exists at all.

Although I've focused on grand goals because they're so exciting, the day-in-day-out progress of science is driven more by little victories. The view from the summit won't sustain a mountain-climber who doesn't like the climb itself. Discoveries in science are relatively rare. The pride of gradual mastery keeps us going in the meantime. Even those theories that don't pan out won't have been a waste. They have opened our minds to new ideas and inspired whole new branches of mathematics.

Given the challenges, it's no surprise that scientists aren't there yet. String theorists readily admit that their field has developed piecemeal and that they don't know what physical principles underlie the theory. The other approaches may have clearer conceptual underpinnings but remain equally mysterious and incomplete. They might well all be wrong.

In the Loop

I believe there is something basic we are all missing, some wrong assumption we are all making. … If string theory or loop quantum gravity by themselves were the answer, we would know it by now. They may be starting points, they may be parts of the answer, they may contain necessary lessons. But the right theory must contain new elements.

—Lee Smolin, Perimeter Institute

Finishing off the job won't be just a matter of solving some more equations and filling in the details. If it were that easy, it'd be done by now. It will require some new conceptual input. The missing pieces may already be out there, scattered among the various alternative approaches like diamonds in the rough, awaiting an Einstein or a Darwin to pull them together. In this regard, it's interesting that the pros and cons of the various approaches to quantum gravity are so complementary. String theory doesn't handle spacetime as fluidly as loop gravity does. Loop gravity doesn't connect as well as string theory does to the observable aspects of spacetime and particles. Some physicists think that each approach may have something to learn from the others.

It may also take some entirely new idea or the recognition that some old idea is subtly flawed. Two places to go hunting for ideas are the nature of time and the status of quantum theory.

What Is Time?

Theories of physics have always had trouble with time. General relativity suggests the world should be playing a giant game of "Freeze!" where nothing moves or changes (see Chapter 7). Quantum theory requires a fixed time standard that a mutable spacetime can't provide (see Chapter 11). These conceptual skeletons in the closet don't affect the day-to-day use of the theories but come tumbling out when physicists try to merge them.

Although Einstein united time with space, that doesn't mean time is the same as space. You can climb a mountain or a tower and see the landscape laid out before you, but you can't climb the temporal equivalent of a mountain or tower and see the future and the past laid before you. You can visit the place of your first kiss as often as you like, but you can visit the moment only once. Time is special. It brings order to the world. Like those psychological tests that ask you to arrange a series of pictures to tell a coherent story, time turns a grab-bag of points into a structured sequence of cause and effect. Without time, there might still be things, sitting in space, but there'd be no stories.

In the Loop

The lack of a proper understanding of time seems to be one of the chief impediments to developing a quantum theory of gravity.

—William Unruh, University of British Columbia

The possibility of time machines is an example of how nature bows down before time. Whenever a time machine seems on the verge of forming, some process intervenes to stop it (see Chapters 10 and 19). And the process that gets in the way varies from situation to situation. It's as though the whole world were acting as time's bodyguards, taking a bullet whenever something threatens the order that time creates.

Physicists routinely imagine universes with fewer or more dimensions of space, but they find it hard to do without time. All the proposed quantum theories of gravity accord time a special status. String theory predicts that the collision of two strings is spread out over space but always occurs at a specific time. Loop gravity, buckyspace, and causal-set theory distinguish between space and time from the outset.

Seen another way, though, time seems so gratuitous. The laws of physics are deterministic—everything that happens, happens for a reason. That's true even for quantum theory, although its determinism hides behind a layer of randomness for reasons that physicists still don't understand. If all events are preordained by the laws of physics, their actual occurrence seems a mere formality. The universe plays itself out as a set piece.

Attempts to unify physics often call determinism into doubt, but violating determinism means violating much else besides. Like the distinction between cause and effect, determinism seems deeply embedded in the structure of the world.

This is just the start of time's mysteries. Is time a product of deeper building blocks? If so, then as time emerges, does the arrow of time—the fact that it seems to flow in one direction only—emerge with it? Could there be places where it flows the other way, as in the movie *Memento* or the novel *Time's Arrow?* A quantum theory of gravity may help solve these questions. Conversely, developing the theory may require physicists to think more deeply about these questions.

Why the Quantum?

All the leading approaches to quantum gravity truly are *quantum* gravity; they take the laws of quantum theory as they are. Those laws are puzzling, but they work. Physicists studying string theory, loop gravity, and the others have never faced a need to abandon or tweak quantum theory. Most are content to leave the conceptual footings of quantum theory to other researchers, who, for their part, rely more on tabletop experiments than on contemplating gravity.

Some physicists, though, think that quantum gravity will transform quantum theory as surely as it does general relativity. Standard quantum theory assumes an unwarped, static, continuous spacetime and might well work differently in a warped, dynamic, or chopped-up one. Broadly speaking, there are three possible outcomes:

- Quantum theory holds in its present form all the way down to the roots of nature, as most physicists now assume.
- Quantum theory holds, but many of its features are specific to the placid, suburban cul-de-sac spacetime we occupy.
- Quantum theory emerges from a deeper theory, maybe hand-in-hand with spacetime itself.

The third option is the most speculative. Physicists taking this route note that quantum theory looks suspiciously like it's hiding something. It says we can't measure all of a particle's traditional properties, such as both its precise position and its precise velocity. In the standard interpretation of the theory, particles don't even possess these properties until we measure them or some equivalent process occurs. Well, why not? Maybe particles do possess such properties and quantum theory just isn't capable of capturing them. A deeper theory could fill in the blanks. Physicists call the properties of such an underlying theory *hidden variables.*

According to the best-known hidden-variables theory, called *Bohmian mechanics* after one of its originators, the quantum wave is a kind of force field herding a particle like a sheepdog. The particle itself behaves completely predictably, but we don't know where it started, so we can't be sure where it'll end up. A bunch of particles that appear to start in the same place with the same velocity actually have a spread of positions and speeds, so they wind up in various places seemingly at random.

In the Loop

Could quantum theory itself be an emergent effective theory? Many have thought so. Extending quantum mechanics until it breaks could be one route to finding out.

—James Hartle, University of California, Santa Barbara

def•i•ni•tion

Hidden variables are quantities that are part of putative theory underlying quantum theory.

Bohmian mechanics is a version of basic quantum theory in which particles have definite positions. Quantum randomness isn't intrinsic to particles but instead reflects our ignorance about them.

Quantum uncertainty reflects our ignorance rather than some built-in element of chance. It's like shuffling a deck of cards, which isn't truly random, but merely very complicated and hard to track.

Early in cosmic history, particles may have been arranged in a more orderly and predictable way, like an unshuffled deck. Since then, they have scrambled themselves. What happened resembled heat death, the concept that the universe is running down as its energy dissipates and its molecules become thoroughly mixed. Quantum theory may be the product of an earlier heat death that occurred at the level of the hidden variables. Our quantum universe is the phoenix that rose from those ashes.

In other hidden-variables theories, the particles we see aren't the true particles but the collective effects of finer building blocks. The randomness has a different source than in Bohmian mechanics. One possible source is spacetime foam—the supposed froth of evanescent black holes at the microscopic level (see Chapter 16). By chewing up information, the holes could throw a wrench into particle processes and make them unpredictable. Another source could be nonlocality—links between points in space (see Chapter 18). Through these links, a particle could get buffeted by events halfway across the universe. If we aren't hip to these links, the particle seems to jiggle around for no good reason. In this case, quantum theory is random because it's procrustean; it chops off those nonlocal links and leaves the loose ends dangling haphazardly. Or the funky behavior of quantum theory may arise because the underlying particles have an unseen dimension of space to play in.

These ideas are controversial. Most physicists think that quantum theory isn't hiding anything. If the theory says a particle doesn't have both a precise velocity and a precise position, that's because the particle doesn't have them, period.

Even if hidden variables don't account for the entire structure of quantum theory, gravitational effects might help explain one puzzling aspect of quantum theory: why particles can come to a fork in the road and take it, whereas people have to choose one path or the other. The randomness arising from nonlocality or spacetime foam might be what forces us to make the choice among the possibilities open to us. This randomness has only a sporadic effect, so a lone particle can evade it for billions of years. But our bodies contain kazillions of particles. Every second, one or another of them gets pinged by nonlocality or spacetime foam. This pinpoints its (and therefore our) location among the multiple possibilities.

Gravity might have other ways of forcing an object to decide among its multiple possible locations. If an object is in two places at once, its two apparitions gravitationally interact. Such an interaction involves energy. Since everything acts to minimize its energy, the system tends to settle into one location or the other. Subatomic particles take a long time to do so because their gravity is so weak, whereas people do so quickly.

If gravity plays a role in making quantum theory what it is, the theory might break down near black holes or in the early universe. Effects now considered forbidden, such as faster-than-light transmission of signals, might become possible. It's a long shot but not crackpot.

For Philosophy

Over the past few years, string theorists have been branching out in directions that are fairly new for them. On the nitty-gritty side, they've sought to make contact with experiment by preparing testable predictions for the Large Hadron Collider, astronomical observatories, and other instruments (see Chapter 21). On the theoretical side, some of them have even sought to make contact with philosophical thinking. Loop theorists have long taken an interest in philosophy, while particle and string theorists have thought of it as woolly. Nonetheless, I've spotted philosophers at recent string conferences.

Historically, the greatest difficulty in scientific revolutions is usually not the missing piece but the extraneous one—the assumption that we've all taken for granted but is actually unnecessary (see Chapter 2). Philosophers are trained to smoke out these mental interlopers. Many of the problems that scientists now face are simply the latest guise of deep questions that have troubled thinkers for thousands of years. Philosophers bring this depth of experience with them. Many have backgrounds in physics as well.

On questions concerning the nature of time and the foundations of quantum theory, academic philosophers have done at least as much work as physicists. They have flagged issues that need to be studied better, such as whether the concept of a law of nature even makes sense. They have also done some thinking about what an ultimate theory will look like. The principles of such a theory will have to differ from the usual sort of explanation, which justifies something in terms of something else. A theory of everything won't have anything else. It will probably have to be self-referential, somehow incorporating its own explanation.

As philosopher Richard Healey of the University of Arizona puts it, philosophers can act as the "intellectual conscience of the practicing physicist." Life would be boring if we always listened to our conscience. Much can be said for serendipity and stumbling across things we didn't expect. Experiment is still the final guide and arbiter. But picking apart the knot of problems at the roots of nature will take a wide variety of intellectual skills.

The Comprehensibility of the Cosmos

The success of physics is in its failure. The basic equations abandon all hope of explaining the world in its full complexity. Physics students learn to understand the laws of motion by temporarily ignoring friction, air resistance, and extenuating circumstances such as different starting points and times. By not getting mired in details, the laws manage to capture the basic essence from which all else emerges.

Remarkably, the world is built in a way that rewards selective ignorance. Physicists often marvel that they can figure anything out at all. They get the distinct impression that ideas await discovery in exactly the same way as a new ice-cream shop on the corner. This feeling is even stronger in string theory than in other theories because it did not grow out of a strong vision. The vision is something to be gleaned by exploring the theory.

The hierarchy of scales, the concept of locality, the distinction between cause and effect, and the principle of symmetry let us understand the universe one piece at a time. We're not confronted with the fearsome task of taking it in all at once. The features that make the universe understandable may be the ones from which it springs to begin with. The fact the universe is comprehensible in pieces will help us to comprehend it as a whole.

The Least You Need to Know

- Quantum gravity is clearly missing some essential elements.
- Some physicists think that the different approaches, such as string theory and loop gravity, might plug one another's gaps.
- Physicists' cluelessness about the true nature of time hampers their efforts to understand quantum gravity.
- Physicists might need to modify or generalize quantum theory to get it to fit with general relativity.

Appendix A

Glossary

annihilation Terrifyingly complete conversion of matter to energy as happens when matter is foolish enough to touch antimatter.

anthropic coincidence Fortuitous value of some quantity in physics, such as the strength of a force, without which our existence would be impossible.

anthropic principle Dictum that we can't formulate a theory or make an observation that is inconsistent with our own existence. It sounds trivial but has profound consequences.

antimatter Material that, if it is brought into contact with ordinary matter, annihilates it in an unhappy burst of radiation. Antimatter looks and acts the same as ordinary matter except that its electric charge, as well as other types of charges such as quark color, is reversed.

antiparticle A particle of antimatter.

arrow of time The distinction between past and future.

atom (1) Smallest possible unit of a chemical element, consisting of a nucleus surrounded by one or more electrons. (2) Smallest possible unit of anything, perhaps including space and time.

background (1) In cosmology, a uniform glow that can't be attributed to specific celestial bodies. (2) In fundamental physics, a fixed framework, such as spacetime, that must be presupposed; it doesn't arise organically from a theory. (3) In experimental science, an overall level of noise that can drown out a signal.

background independence Concept that spacetime is not a fixed framework but can change shape.

big bang Expansion of the universe from a hot, dense state. I use the term to refer to the expansion, which is still taking place. Scientists sometimes use the term to indicate just the moment when the expansion started, the putative beginning of time.

black hole Tightest possible wadding of matter, surrounded by an event horizon.

black-hole information paradox The prospect that information falling into a black hole gets destroyed, contradicting quantum theory.

Bohmian mechanics Version of basic quantum theory in which quantum randomness isn't intrinsic to particles but, instead, reflects our ignorance about them.

bootstrap paradox A paradox of time travel in which an event causes itself or an object creates itself. The idea was made famous by science-fiction writer Robert Heinlein's creepy story *All You Zombies*.

boson Type of particle that can clump with identical copies of itself like a clone army. Particles of force fall into this category.

brane Short for membrane. A two-dimensional sheet or analogous object in lower or higher dimensions as hypothesized by string theory. Pronounced the same as brain.

braneworld D-brane or set of D-branes on which a universe can reside. Our observable universe may be one.

brute fact Fact with no deeper explanation. It just is.

buckyspace My unofficial name for causal dynamical triangulations; the inventors of the theory don't seem to mind.

Calabi-Yau shape Funky shape into which the extra-dimensional space of string theory may be crumpled up.

causal dynamical triangulations A quantum theory of gravity that approximates space as a scaffolding of triangles, each designed to ensure that cause always precedes effect.

causal set Network of events organized according to which caused which.

CERN World's premier particle physics lab, located at the western end of the bus line (soon to be tram line) in Geneva, Switzerland, straddling the border with France. In full: *Conseil Européen pour la Recherche Nucléaire*.

charge Property of a particle that determines how strongly it interacts with other particles. Electric charge is an example.

classical In physics, an adjective referring to a prequantum theory. Einstein's theories of relativity and Maxwell's equations are classical theories. They deal in quantities such as position and velocity that always have definite values.

closed string String that forms a loop like a miniature rubber band.

color (1) In ordinary life, a property associated with the wavelength of light. (2) In fundamental physics, the charge associated with the strong nuclear force. It got its name because it comes in three types, like the three primary colors.

Compton wavelength Squeezing a particle to this wavelength gives it enough energy to create a whole new particle.

conservation of angular momentum Principle that an isolated object doesn't gain or lose angular momentum, the momentum associated with rotational motion, over time.

conservation of energy Principle that an isolated object or collection of objects doesn't gain or lose energy over time.

conservation of momentum Principle that an isolated object doesn't gain or lose momentum over time.

coordinate Number that identifies the position of an object or event.

cosmic censorship conjecture Hypothesis that naked singularities are impossible. Whenever matter gets wadded tightly enough, an event horizon forms to hide it from our prying eyes.

cosmic coincidence Chill-up-your-spine fact that dark energy became dominant around the same time that stars and galaxies formed.

cosmic expansion Tendency of space to stretch, pulling galaxies apart.

cosmic microwave background radiation Afterglow of the early stage of the big bang. Consists of photons that once interacted with matter, until parting ways about 400,000 years after the start of the bang. Also called *cosmic microwave background* or simply *microwave background.* Abbreviated CMBR.

cosmic rays Not really rays, but particles zipping through outer space at high speed.

cosmic string Long filament of energy. It could be a kink in quantum fields or a string-theory string enlarged by cosmic expansion.

cosmological constant Form of energy that pervades spacetime. It may arise from quantum fluctuations within spacetime or be woven into the fabric of spacetime itself. It's one of the possible explanations for dark energy.

cosmological constant problem The strange fact that quantum theory predicts a much larger value for the cosmological constant than astronomers observe.

cosmology Subbranch of astronomy that studies the origin and overall evolution of the universe.

curvature In fundamental physics, the degree to which spacetime is bent. It varies from place to place depending on the local density of matter and energy. It's associated with gravity.

dark energy Unknown type of energy that seems to be causing cosmic expansion to accelerate, pulling galaxies apart at ever-increasing speeds. Not to be confused with dark matter.

dark matter Unknown material that makes up the bulk of the matter in the universe. It neither emits nor absorbs light but exerts a gravitational force and possibly the weak nuclear force. Not to be confused with dark energy.

decay Spontaneous disintegration of a particle into two or more other particles.

decoherence Process that causes quantum objects to lose their distinctively quantum features and behave like run-of-the-mill classical objects.

determinism Concept that the universe follows a specific course of behavior. Its condition at any one moment dictates what will happen in the future and what must have happened in the past.

dilaton Quantum field that governs the strength of string interactions. In string theory, it acts as master control determining how powerful the forces of nature are.

dimension An independent direction of possible motion in space or time. The idea can be generalized to other properties of space, in which case the number of dimensions doesn't have to be a whole number. In some uses of the term, space is not ordinary space but a more abstract realm of possibilities.

discrete Spaced out at intervals, like rungs on a ladder. Opposite of continuous.

doublet Pair of particles related by the weak nuclear force, such as the electron and electron neutrino.

doubly special relativity Modified version of Einstein's special theory of relativity with two universal limits, the speed of light and the Planck scale. Also called *deformed special relativity*.

duality Principle that connects two seemingly distinct situations and shows they are actually one and the same. Much beloved by string theorists.

ekpyrosis Hypothesis that the big bang began in a blaze when our universe smashed into a parallel one.

electromagnetic field Field that gives rise to electromagnetic radiation.

electromagnetic radiation Interlocking, self-reinforcing oscillations of electric and magnetic fields. It encompasses, in order of ascending energy content, radio waves, infrared light, visible light, ultraviolet light, x-rays, and gamma rays.

electromagnetism Union of electricity and magnetism.

electron A friendly, lightweight, negatively charged elementary particle.

electron-volt The standard unit of energy in particle physics and string theory. It's the energy gained by an electron accelerated by a voltage difference of one volt. Abbreviated eV. It takes standard metric prefixes: milli, kilo, mega, giga, tera, and so on.

electroweak forces Union of the electromagnetic and weak nuclear forces.

electroweak theory Theory describing the electroweak forces. It generalizes quantum electrodynamics.

elementary particle Fundamental building block, with no internal structure. The electron is one. The proton is not.

emergence Process whereby a complex system acquires properties that its components don't have.

energy Property of an object responsible for bringing about change and producing the force of gravitation. It comes in various forms, such as kinetic energy (associated with overall motion of a body), potential energy (associated with position within spacetime or a more abstract realm), and thermal energy (heat content). Mass can be thought of as a form of energy built into objects.

entanglement Quantum version of a romantically complicated situation. Two particles establish a connection and maintain it no matter how far apart they move.

entropy A measure of how many ways molecules can be arranged for a given amount of energy. It's the collective information content of the molecules.

equivalence principle Concept that the acceleration of an object is indistinguishable from artificial gravity. This principle, the basis of Einstein's general theory of relativity, implies that the laws of physics work the same for all observers as long as they're smart enough to account for the gravitational field.

ESA European Space Agency, the European counterpart to NASA.

eternal inflation Hypothesis that cosmological inflation, once started, continues forever in the universe at large. The observable universe is just a small region where it happened to pause or stop.

event Specific position in space at a specific time. It's the basic unit of spacetime.

event horizon Perimeter of a black hole, marking the point of no return for infalling material. Roaches check in but they don't check out. Those of us safely ensconced on the outside can't see the carnage inside.

extra dimensions Dimensions of space beyond the usual three of length, width, and height. If these dimensions exist, they're hidden from us, at least for now.

Fermilab Fermi National Accelerator Laboratory, located west of Chicago, is the home to the Tevatron and one of the world's leading centers for particle physics.

fermion Type of particle that stays clear of identical copies of itself. Particles of matter fall into this category.

field In fundamental physics, a type of substance that fills space like a fog. It's described by one or more numbers at each point in space and time. Examples include the electric field, magnetic field, and gravitational field.

force Effect that acts to change the motion of a body, either its speed or direction or both.

fractal Geometric figure that repeats its overall shape on multiple scales.

galaxy Giant system of stars, gas, dust, and dark matter. In essence, it's the next level of organization up from the solar system. Galaxies are large enough and widely separated enough to feel the effects of cosmic expansion.

gamma rays Most energetic type of electromagnetic radiation, even more penetrating than x-rays.

gauge symmetry *See* local symmetry.

general theory of relativity Einstein's theory of gravitation, which attributes gravitational forces to the curvature of the spacetime continuum. Also called *general relativity* for short.

generation One of the periodic groups of matter particles in the Standard Model. Also called *family.*

giga Prefix meaning billion (a thousand million).

GLAST Gamma-Ray Large Area Space Telescope, a robotic orbital observatory built by an international collaboration to study high-energy gamma radiation.

global symmetry Type of symmetry in which the transformation, such as rotation, acts equally everywhere.

gluon Particle that transmits the strong nuclear force. It glues quarks together to form protons and neutrons.

Grand Unified Theory Theory that describes electromagnetism, the weak nuclear force, and the strong nuclear force as aspects of a single primordial force. It does *not* include gravity. Abbreviated GUT.

grandfather paradox The classic paradox of time travel, in which a time traveler kills his or her own grandfather or other forbear, thereby preventing himself or herself from ever being born.

gravitation Force generated by mass or energy. It reflects the warping of spacetime. It usually creates an attraction, but special forms of energy produce a repulsion. This book uses the terms gravitation and gravity interchangeably.

gravitational wave Interlocking, self-reinforcing oscillations of gravitational forces. It's a ripple in the gravitational field and, therefore, in the fabric of spacetime. Careful: it's *not* the same as a gravity wave, a phrase reserved for a separate concept in atmospheric science. Also called *gravitational radiation.*

graviton Hypothesized particle that transmits the force of gravity. It's the smallest unit of a gravitational wave.

gravity/gauge duality Principle that a universe with gravity is completely equivalent to a lower-dimensional universe without gravity. Also known as AdS/CFT correspondence and the Maldacena conjecture after the scientist who first proposed it.

hadron Subatomic particle made up of two or more quarks or antiquarks; examples include protons and neutrons.

Hagedorn temperature In string theory, a special temperature that is either the highest temperature strings can attain or a type of boiling point between strings and black holes.

Hawking effect Shrinkage of a black hole as particles escape from its perimeter. Also called *Hawking radiation.*

hidden variables Quantities that are part of a putative theory underlying quantum theory.

hierarchy problem An awkward disparity of scales. Processes occur at vastly different scales when, by all rights, they should occur at the same scale.

Higgs field Hypothetical quantum field, with associated particle, that's thought to be responsible for giving other elementary particles their mass. It's a scalar field, a primitive substance different from other ingredients of the Standard Model.

Higgs hierarchy problem Strange fact that the Higgs particle has a low mass even though the particles affecting it have a high mass.

hologram Special type of photograph that captures a 3-D scene with depth.

holographic principle Concept that the amount of information in a region is proportional not to its volume but to the area of its boundary.

horizon problem Strange fact that widely separated regions of our universe are nearly identical despite the distance between them.

hypercube Four-dimensional cube. Also called a *tesseract.*

hyperpyramid Four-dimensional pyramid.

hypersphere Four-dimensional sphere.

indeterminism Opposite of determinism. Two identical situations can have the same outcome or, conversely, two outcomes could have had exactly the same cause.

infinitesimal Having zero extent. Infinitely small.

inflation Of the cosmological variety, an acceleration of the rate of cosmic expansion, pulling regions of space apart too fast for them to exchange signals or other influences. Widely thought to have occurred very early in the history of our universe. The present cosmic acceleration may be a recurrence.

inflaton Hypothesized field and associated particle that brought about inflation. It is a scalar field, like the Higgs field. The word looks like a typo of "inflation," but it is correct as you see it.

information paradox *See* black-hole information paradox.

infrared Light just beyond the red end of the visible spectrum. It's less energetic than visible light but more so than radio waves.

interaction Generalization of the concept of force. It occurs when two or more particles swap a boson, leading to a change in their velocity or other properties.

interference Overlap and intermingling of waves, one of the defining features of quantum theory.

internal symmetry Principle that the laws of physics don't change even when internal properties of particles do.

kaon Type of particle that consists of two quarks. It's the canary in the coal mine of physics, peculiarly sensitive to new physical effects.

kelvin Standard unit of temperature equaling 1 degree Celsius or 1.8 degrees Fahrenheit. A temperature of 0 kelvins is absolute zero, the lowest possible temperature. 0°C (32°F) corresponds to 273.15 kelvins. Note: for consistency with other units of physics, don't say "degrees kelvin."

kinetic energy Energy associated with motion.

landscape In string theory, the range of possible shapes that the unseen dimensions of space could have.

Large Hadron Collider World's highest-energy particle accelerator, located at CERN, which began operation in 2008. Abbreviated LHC.

laser-ranger Laser used to measure distance precisely.

left-handed One of the two mirror images of a particle, related to how it responds to the weak nuclear force.

lepton Category of particle that includes electrons and neutrinos. It's immune to the strong nuclear force, so it doesn't get trapped in atomic nuclei.

LIGO Laser Interferometer Gravitational Wave Observatory, with stations in eastern Washington state and southern Louisiana.

local symmetry Symmetry in which the transformation can act differently at different points in space. Also called *gauge symmetry.*

locality Concept that what happens in one place doesn't directly affect another. Something must pass between the two places to carry the influence.

loop quantum gravity Leading alternative to string theory for a quantum theory of gravity. It describes space in terms of linked atoms of volume. The eponymous loops are an abstract feature. Also called *loop gravity.*

M-theory Theory that is thought to underlie string theory. Its details are mysterious.

macroscopic Large. Big, at least to a particle physicist. Refers to objects we can see directly, from a speck of dust on up.

many-worlds interpretation Way of making sense of quantum randomness in terms of the existence of parallel universes.

mass Energy built into an object, causing it to resist being accelerated. Also called *rest mass.*

matter Stuff from which the universe is made. Made up of fermions such as quarks, leptons, and dark matter.

Maxwell's equations Unified prequantum description of electromagnetism. A prototype of a unified theory of physics supplanted by quantum electrodynamics.

model In theoretical physics, an abstract representation of the real world. Usually considered one notch down from a full-fledged theory, in that it is incomplete and typically describes rather than explains.

Moiré pattern Wavy pattern that results from laying one fine-scale pattern on top of another.

momentum Quantity representing the power of a moving body and the difficulty of stopping it. A massive body moving slowly has the same momentum as a lighter body moving faster. A light beam has zero mass, but still has momentum since it's going so fast.

multiverse Larger volume of space of which our observed universe is only a part.

naked singularity Object resembling a black hole except that it lacks an event horizon.

nano Prefix meaning billionth.

NASA National Aeronautics and Space Administration, the U.S. government space agency.

negative energy (1) Amount of energy less than zero—that is, less than what the vacuum contains. Quantum theory predicts it can exist for limited periods of time. Also called *exotic matter.* (2) Amount of energy less than some reference level, sometimes defined as the amount necessary to hold a body together. This definition is really just a matter of convention, whereas the first definition involves a qualitatively new phenomenon.

neutralino Hypothetical supersymmetric particle thought to be the lightest possible. It might be the universe's dark matter. Not to be confused with neutrino.

neutrino Lightest known particle, apart from particles with no mass at all. It has no electric charge and barely interacts with other forms of matter, but is the linchpin of nuclear reactions. Not to be confused with neutralino.

neutron Electrically neutral particle found in atomic nuclei. It consists of three quarks, two down, one up.

Newton's law of gravity Classic equation for the force of gravity, in which gravity weakens in inverse proportion to the square of the distance. It has been supplanted by relativity theory but remains a good approximation.

no-hair theorem Principle that a black hole is an utterly featureless body characterized fully by its mass, charge, and spin. It's a prediction of relativity theory.

noncommutative geometry Properties of a spacetime where a zero-sum game applies to the coordinates of each point. The more precise one coordinate is, the less precise the others can be.

nonlocality Concept that objects at widely different locations are still somehow connected, even if nothing tangible stretches between them.

nucleus Core of an atom, comprising protons and neutrons.

observable universe Sum of all we can see. It may be just one part of the full reality. Also called *our universe.*

Olbers' paradox Observation (not really a paradox) that if the universe were infinitely old, the night sky would be totally plastered with stars and therefore bright.

open string String that's an open-ended filament.

order of magnitude (1) Factor of 10. (2) Value rounded off to the nearest power of 10. The sort of approximate number we get from a back-of-envelope calculation.

particle Building block of nature that can be either elementary (having no internal structure: an electron) or composite (having an internal structure: a proton). Quantum theory describes particles as the smallest unit of oscillation of their respective field.

particle accelerator Device to propel particles to high speed. An ordinary battery is actually a weak particle accelerator; it propels electrons through a wire. The cathode ray tube in an old-style TV set is one, too. Physicists use accelerators to smash particles together to create new ones. Also called a *particle collider* or *atom smasher.*

phase transition Change of matter from one state to another. Boiling and freezing are examples.

photon Particle that transmits electromagnetic forces and makes up electromagnetic radiation.

pico Prefix meaning trillionth.

Planck energy Energy at which the quantum nature of gravity becomes apparent and thought to be the highest possible energy an elementary particle can have. Numerically equal to about 1.2×10^{19} GeV. Roughly the same as the chemical energy in a tank of gasoline, concentrated into a single particle.

Planck length Distance at which the quantum nature of gravity becomes apparent. It's thought to be the smallest possible length, numerically equal to about 1.6×10^{-35} meter. About as small in relation to an atom as a human being is to the observable universe.

Planck mass Mass corresponding to the Planck energy, according to Einstein's famous equation $E = mc^2$. It's thought to be the largest possible mass an elementary particle can have; also the smallest mass a black hole can have, unless gravity unexpectedly strengthens on small scales. Numerically equal to about 20 micrograms, roughly the mass of a paramecium.

Planck scale Defining scale of quantum gravity, which can be expressed either as energy, length, mass, temperature, or time—they're all equivalent.

Planck Surveyor New ESA satellite to measure the cosmic microwave background radiation.

Planck temperature Temperature of a gas whose particles have the Planck energy. It's thought to be the highest possible temperature, numerically equal to about 1.4×10^{32} kelvins.

Planck time Time it takes for a light beam to cross the Planck length. It's thought to be the smallest meaningful time interval, numerically equal to about 5×10^{-44} second. Smaller in relation to a picosecond than a picosecond is to the age of our universe.

point Dot of zero size.

polarization For waves, the direction the waves oscillate in.

positron Antimatter counterpart of the electron.

pre–big bang scenario Cosmological model based on string theory, in which time extends back before the start of the big bang.

principle of relativity Concept, introduced by Galileo Galilei, that you can't tell whether you or a reference point is moving. It implies that the laws of physics work the same for all observers. Einstein adopted it as the basis of his theories.

problem of frozen time Conundrum that according to the general theory of relativity, the world should be static and unchanging. Also called the *problem of time*.

proton Positively charged particle found in atomic nuclei or on its own, consisting of three quarks, two up and one down.

quantum (1) Smallest unit into which something can be subdivided. A particle is the quantum of its respective field. Plural: quanta. (2) Anything that obeys the principles of quantum theory.

quantum field theory Theory that unifies quantum mechanics with the special theory of relativity. It describes particles as the quanta of fields.

quantum fluctuation Random variation in quantities resulting from quantum effects.

quantum leap (1) Abrupt jump in energy that occurs when an atom absorbs one quantum. It's small, unlike the common usage of the term. (2) Unaccountably random result obtained when measuring a quantum system.

quantum mechanics Theory of the behavior of objects on the assumption that matter, energy, and force come in indivisible units. Technically, quantum mechanics, as opposed to quantum field theory, refers to a finite number of discrete objects.

quantum theory General term encompassing quantum mechanics and quantum field theory in their various incarnations.

quantum theory of gravity Type of theory of which string theory and loop quantum gravity are examples. It describes the force of gravitation in quantum terms, thereby unifying quantum theory with the general theory of relativity. Also called *quantum gravity* for short.

quark Category of particle that makes up protons and neutrons. It feels all the known forces of nature, including the strong nuclear force.

radiation Not just the dangerous stuff given off by nuclear fallout but also any form of propagating energy, including light, which is electromagnetic radiation, and gravitational waves.

reductionism Principle that complex phenomena can be broken down into smaller pieces that are easier to explain. Often taken to mean that the laws governing the smaller pieces suffice to describe the universe fully.

relativity theory Einstein's theories of motion and gravity, encompassing the special theory of relativity and the general theory of relativity.

renormalizability Type of scale-invariance of particle processes. If a force of nature is renormalizable, the basic template of reactions involving this force is the same at all scales.

right-handed One of the two mirror images of a particle, related to how it responds to the weak nuclear force.

scalar field Special type of field that can be described by a single number at each point in space, as opposed to both a number and a direction. Temperature and pressure are everyday examples. The Higgs field and inflaton are physics examples.

scale-invariance Property of a process or object whereby it works the same no matter what size it is.

Schrödinger's cat Thankfully hypothetical kitten that is both alive and dead at the same time. Quantum theory allows for such seemingly contradictory situations.

selectron Supersymmetric partner of the electron. It's a scalar particle.

singularity Location in spacetime where a theory predicts that quantities such as density become infinite. Centers of black holes are an example.

space Not just outer space but the space all around us. Very loosely speaking, it's the container in which objects live and play.

space invader Not just a classic video game, but an object that can reach us from infinitely far away. Such objects would wreck the predictive power of theories.

spacetime Union of space and time, as introduced by Einstein.

spacetime foam Frothy behavior of spacetime on the smallest scales, caused by quantum fluctuations of the gravitational field. May include the appearance and disappearance of tiny black holes and wormholes.

sparticle Hypothetical supersymmetric partner of a Standard Model fermion.

special theory of relativity Einstein's theory of motion in the absence of gravity. Holds that the speed of light is the same for all observers, that space and time are interrelated, and that mass is a form of energy, according to $E = mc^2$. Even when gravity operates, the theory still applies approximately over small regions of spacetime. Also called *special relativity.*

speed of light Maximum speed not only of light but also of any object or process. Numerically equal to about 300,000 kilometers per second. Denoted by c.

spin In quantum theory, the analogue of rotational motion. It governs whether particles are bosons or fermions.

spin network Connect-the-dots diagram showing the relationships among events.

squark Hypothetical supersymmetric partner of the quark. It's a scalar particle.

Standard Model Current explanation for the composition of the material world and the operation of the electromagnetic, strong nuclear force, and weak nuclear force.

string Vibrating band that may be the fundamental building block of nature. It comes in different varieties, including the superstring, which incorporates supersymmetry.

string theory Quantum theory of strings, considered the leading candidate for a grand unified theory and a quantum theory of gravity. One version, known as *superstring theory*, incorporates supersymmetry.

strong nuclear force Force responsible for binding quarks and atomic nuclei together. Also called *strong force* or *strong interaction.*

Super-Kamiokande Particle detector based on 50,000 tons of water, used to look for neutrinos and decaying protons and located in central Japan west of Tokyo.

supergravity Union of supersymmetry with the general theory of relativity. Though incomplete as a quantum theory of gravity, it serves as a useful approximation to string theory.

supermassive black hole Black hole with too much mass to have been created by the collapse of a single star. Thought to arise from the merger of smaller holes or collapse of primordial gas clouds, it resides in the core of a galaxy.

superposition In quantum theory, a mixture of possible outcomes.

supersymmetry Principle that fermions are related to bosons.

symmetry Principle that the properties of objects or of equations don't change even when they are transformed in some way.

symmetry-breaking Process whereby a symmetrical situation becomes asymmetrical.

T-duality In string theory, the equivalence between a small dimension of space and a large one.

tera Prefix meaning trillion.

Tevatron Particle accelerator at Fermilab. The record-holder for energy until the Large Hadron Collider came online.

theory Well-developed conceptual framework.

theory of everything Tongue-in-cheek term for a unified theory of all particles and forces.

time dilation Principle, derived from Einstein's special theory of relativity, according to which time passes more slowly for a moving object than for a stationary one.

torus Doughnut shape.

translation In physics, picking up and moving something from one place or time to another without rotating it.

ultraviolet Light just beyond the violet end of the visible spectrum. It's more energetic than visible light but less so than x-rays.

uncertainty principle Quantum zero-sum principle. Certain properties of nature come in pairs, and if one of the properties fluctuates less, the other must fluctuate more. Also called *Heisenberg's uncertainty principle.*

unified theory Single set of underlying principles that describes outwardly distinct forces and types of matter.

universality Principle that different systems can have the same behavior under circumstances such as phase transitions. Compositional details cease to matter.

vacuum Lowest-energy stable state of space. In quantum theory, it isn't completely empty; the fields are still there, merely dormant.

velocity Speed and direction of an object's motion.

virtual particle Distinctive type of particle described by quantum theory. It exists for too short a time to be detected directly, can convey forces, and is a way to describe the irrepressible fluctuations of quantum fields.

W boson One of the particles that transmits the weak nuclear force and can transmute particles.

wavefunction Description of the location of a quantum particle or system in terms of probabilities.

weak nuclear force Interaction responsible for transforming electrons into neutrinos and up quarks into down quarks, among other effects. Also called *weak force* or *weak interaction.*

winding Ability of a string to wrap around space.

worldsheet Two-dimensional surface that strings trace out as they move through spacetime. Its definition is the mathematical heart of string theory.

wormhole Alternate path between two points in spacetime.

Z boson One of the particles that transmits the weak nuclear force. Exerting a force similar to that of the photon, but much more limited in range, it's said to have been given its name because physicists thought it was the last particle they'd need to discover. How wrong they were!

Appendix B

Selected Readings

Visit www.strings.musser.com for book errata and links to other resources.

The best way to really learn about science is to take a class. That way, you have someone to turn to with questions, rather than trying to puzzle it out on your own. As for Internet discussion boards and blogs, sorting the useful from the distracting can be very hard. One rule of thumb is to beware of overconfidence. Those who forthrightly acknowledge the limits of their own knowledge are often the ones who know the most. Watch out for asymmetry. Do people apply their criticism to another theory but not to their own? Or do they shoot down a theory without offering any alternative? A respectful yet penetrating question can be more revealing than a broadside attack.

Books

About Physics in General

Bronowski, Jacob. *Science and Human Values*. New York: Harper & Row, 1956. This is an extended essay, one of the best, on what science means.

Davies, Paul, ed. *The New Physics*. New York: Cambridge, 1989. If you want to go in depth, go here.

Einstein, Albert. *Out of My Later Years.* Secaucus, NJ: Citadel, 1995. A collection of the master's essays, notably "Physics and Reality" from 1936, which is the source of the famous line, "The eternal mystery of the world is its comprehensibility."

Gamow, George. *Mr. Tompkins in Paperback*. New York: Cambridge, 1965. This oldie but goodie explains what the world would look like if the speed of light were 10 miles per hour and pool balls behaved like quantum particles.

Weinberg, Steven. *Dreams of a Final Theory*. New York: Vintage Books, 1992. Here is another extended essay on why physicists do what they do, as well as the author's thoughts on religion and philosophy. Unfortunately, its eloquence couldn't convince Congress to continue funding the Superconducting Super Collider.

About the Standard Model

Feynman, Richard P. *QED: The Strange Theory of Light and Matter*. Princeton, NJ: Princeton, 1985. The famous iconoclast who used to pick locks for fun during the Manhattan Project gives the definitive explanation of the core of the Standard Model.

Quinn, Helen R., and Yossi Nir. *The Mystery of the Missing Antimatter.* Princeton, NJ: Princeton, 2008. The most up-to-date discussion of why matter and antimatter are so out of whack and what this means for the unification of physics.

About Time Travel

Davies, Paul. *About Time: Einstein's Unfinished Revolution.* New York: Simon & Schuster, 1995. By the end, you'll marvel at how something so basic as time can be so poorly understood.

Gott III, J. Richard. *Time Travel in Einstein's Universe: The Physical Possibilities of Travel Through Time*. New York: Houghton Mifflin, 2001. Here is an entertaining account of one way to build a time machine.

Thorne, Kip S. *Black Holes and Time Warps: Einstein's Outrageous Legacy.* New York: W.W. Norton, 1994. This is a must-read account of how a wormhole time machine might work, from the physicist who worked it out—not to mention charming sketches and dramatic tales of private cab rides in Soviet-era Moscow.

About Cosmology

Bartusiak, Marcia. *Through a Universe Darkly: A Cosmic Tale of Ancient Ethers, Dark Matter, and the Fate of the Universe*. New York: HarperCollins, 1993. This book brings dark matter down to Earth, in more ways than one.

Borges, Jorge Luis. *Ficciones*. Buenos Aires: Emecé Editores, 1956. Here find short yet profound stories about topics such as the multiverse, including "La Biblioteca de Babel" ("The Library of Babel"). Online at jubal.westnet.com/hyperdiscordia/library_of_babel.html

Harrison, Edward R. *Cosmology: The Science of the Universe*. New York: Cambridge, 2000. This textbook that doesn't read like one is for those who want to dig deeper.

Hogan, Craig J. *The Little Book of the Big Bang: A Cosmic Primer*. New York: Springer-Verlag, 1998. The title says it all.

Rees, Martin. *Before the Beginning: Our Universe and Others*. Reading, MA: Addison-Wesley, 1997. This eminent astronomer explores the idea of a multiverse and how it might be observed.

Steinhardt, Paul J., and Neil Turok. *Endless Universe: Beyond the Big Bang*. New York: Doubleday, 2007. The string-theoretic version of the cyclic universe is presented by two of the physicists who worked it out.

About Quantum Theories of Gravity

Greene, Brian. *The Elegant Universe: Superstrings, Hidden Dimensions, and the Quest for the Ultimate Theory*. New York: Vintage Books, 1999. This classic account of string theory, if now a bit dated, gives great explanations of basic concepts and tales of what it's like to do science at the cutting edge.

Kane, Gordon. *Supersymmetry: Unveiling the Ultimate Laws of Nature*. Cambridge, MA: Perseus Publishing, 2001. This work focuses on the one aspect of string theory, supersymmetry, that will probably live on regardless of what happens to the full theory.

Randall, Lisa. *Warped Passages: Unraveling the Mysteries of the Universe's Hidden Dimensions*. New York: HarperCollins, 2005. Somewhat more technical than Greene's book, this is still engaging and is an especially good resource for understanding branes.

Smolin, Lee. *Three Roads to Quantum Gravity*. New York: Basic Books, 2001. More philosophical than Greene's book, this is the best place to go for explanations of the leading alternatives to string theory.

Susskind, Leonard. *The Cosmic Landscape: String Theory and the Illusion of Intelligent Design*. New York: Little, Brown & Co., 2006. One of string theory's founding fathers explains the landscape concept and much besides. Few scientists have such a strong writing style; you can practically hear Susskind talking to you.

About Extra Dimensions

Abbott, Edwin A. *Flatland: A Romance of Many Dimensions*. Princeton, NJ: Princeton, 1991. This is a reissue of the nineteenth-century classic tale of the square who meets the sphere, with an introduction by mathematician Thomas Banchoff, the world's leading expert on visualizing extra dimensions. Original text is online at www.gutenberg.org/etext/97.

Dewdney, A. K. *The Planiverse: Computer Contact with a Two-Dimensional World*. New York: Simon & Schuster, 1984. The book describes how a flat universe might really work, right down to a creature's intestinal tract.

Rucker, Rudy. *The Fourth Dimension: Toward a Geometry of Higher Reality*. New York: Houghton-Mifflin, 1984. This is a fun book with quirky diagrams to guide you through dimensional mathematics and physics.

About Experiments

Bartusiak, Marcia. *Einstein's Unfinished Symphony: Listening to the Sounds of Space-Time*. Washington, D.C.: Joseph Henry Press, 2000. The book explains the search for gravitational waves and the epic technological advances it has required.

Articles

About Physics in General

Matthews, Robert A. J. "Tumbling toast, Murphy's Law and the fundamental constants," *European Journal of Physics*, Vol. 16, (1995); 172–176. This gives the technical account of why toast falls with the buttered side down, as mentioned in Chapters 2 and 15.

About Quantum Theory

Kwiat, Paul G., and Lucien Hardy. "The mystery of the quantum cakes," *American Journal of Physics*, January 2000. This version of Bell's experiments inspired the second shell game in Chapter 4.

Mermin, N. David. "Is the Moon There When Nobody Looks? Reality and the Quantum Theory," *Physics Today*, April 1985. Here is the definitive explanation of Bell's experiments, which inspired the first shell game in Chapter 4.

Pesic, Peter. "Quantum Identity," *American Scientist*, May/June 2002. This article builds up quantum theory from the fact you can't tell one electron from another.

Tegmark, Max, and John Archibald Wheeler. "100 Years of Quantum Mysteries," *Scientific American*, February 2001. A rising star and a physics veteran explain what quantum theory has done and where it's going.

About the Standard Model

Dimopoulos, Savas, Stuart A. Raby, and Frank Wilczek. "Unification of Couplings," *Physics Today*, October 1991. Here are the details of how the three forces of nature might become one.

Kane, Gordon. "The Dawn of Physics Beyond the Standard Model," *Scientific American*, June 2003. This gives you a to-do list for particle physics.

About Black Holes and Wormholes

Carr, Bernard J., and Steven B. Giddings. "Quantum Black Holes," *Scientific American*, May 2005. Online at tinyurl.com/34u36j. This explains how you can make a black hole in the lab and why you'd ever want to.

Ford, Lawrence H., and Thomas A. Roman. "Negative Energy, Wormholes, and Warp Drive," *Scientific American*, January 2000. This article inspired so many letters to the editor that we had to haul in a crate to hold them all.

Susskind, Leonard. "Black Holes and the Information Paradox," *Scientific American*, April 1997. What happens if you throw the *Encyclopedia Britannica* into a black hole?

About Cosmology

Holt, Jim. "Nothing Ventured," *Harper's Magazine*, November 1994. Leading writer on the philosophy of science asks why there is something rather than nothing.

Lineweaver, Charles H., and Tamara M. Davis. "Misconceptions About the Big Bang," *Scientific American*, March 2005. Even the experts sometimes get it wrong.

Primack, Joel R., and Nancy Ellen Abrams. "In a Beginning … Quantum Cosmology and Kabbalah," *Tikkun*, January/February 1995. Here you find modern cosmology in terms of metaphors from Jewish mysticism. You don't need to be Jewish or a mystic to find it insightful. Online at physics.ucsc.edu/cosmo/primack_abrams/InABeginningTikkun1995.pdf

Tegmark, Max. "Parallel Universes," *Scientific American*, May 2003. One of the most popular articles the magazine ever published, this was selected for *Best American Science & Nature Writing*.

Veneziano, Gabriele. "The Myth of the Beginning of Time," *Scientific American*, May 2004. The big bang was the outcome of something rather than the cause of everything.

About Quantum Theories of Gravity

Arkani-Hamed, Nima, Sava Dimopoulos, and Georgi Dvali. "The Universe's Unseen Dimensions," *Scientific American*, August 2000. Is gravity leaking out of our universe into a higher-dimensional realm?

Bartusiak, Marcia. "Loops of Space," *Discover*, April 1993. Wherein a scientist scours the shops of Verona, Italy, for key rings.

Bekenstein, Jacob D. "Information in the Holographic Universe," *Scientific American*, August 2003. Pioneer of the quantum theory of black holes describes why 3-D space may be a grand illusion.

Chalmers, Matthew. "Stringscape," *Physics World*, September 2007. Online at physicsworld.com/cws/article/print/30940. This is probably the best single article on string theory.

Galison, Peter. "Theory Bound and Unbound: Superstrings and Experimeter," *Laws of Nature*, eds. Weinert, Friedel, and Walter de Gruyter, 1995. This academic article by the renowned historian of science is the single best account so far of the early history of string theory and the String Wars.

Maldacena, Juan. "The Illusion of Gravity," *Scientific American*, November 2005. This definition account of duality is from the man who started it all, not to mention inspired a late-'90s dance craze in physics.

Smolin, Lee. "Atoms of Space and Time," *Scientific American*, January 2004. One of the fathers of loop quantum gravity spells it out.

About Experiments

Barrow, John D., and John K. Webb. "Inconstant Constants," *Scientific American*, June 2005. This gives controversial hints that the constants of nature aren't.

Cline, David B. "The Search for Dark Matter," *Scientific American*, March 2003. Dark particles are streaming through our bodies all the time.

Gibbs, W. Wayt. "Ripples in Spacetime," *Scientific American*, April 2002. Will loblolly pine trees drown out the sounds of spacetime?

Hedman, Matthew. "Polarization of the Cosmic Microwave Background," *American Scientist*, May/June 2002. Is grand unification written on the sky?

Kane, Gordon. "String Theory Is Testable, Even Supertestable," *Physics Today*, February 1997. Technical version online at arXiv.org/abs/hep-ph/9709318. Article explains how string theory connects to the real world.

Kolbert, Elizabeth. "Crash Course," *New Yorker*, May 14, 2007. Online at www.newyorker.com/reporting/2007/05/14/070514fa_fact_kolbert. Physicists get ready for the Large Hadron Collider by swigging one espresso coffee after another.

Smith, Chris Llewellyn. "The Large Hadron Collider," *Scientific American*, July 2000. Former director of CERN lab describes how the discovery machine works. At the time, they thought the LHC would start up in 2005. Better late than never.

Websites

About Physics in General

How to become a theoretical physicist, by Nobel laureate Gerard 't Hooft: www.phys.uu.nl/~thooft/theorist.html#stheory

The "baloney detection kit," a big help in sorting good websites from bad: homepages.wmich.edu/~korista/baloney.html

The official physics FAQ:
www.math.ucr.edu/home/baez/physics

About Relativity Theory

Relativity tutorial, by renowned philosopher John D. Norton:
www.pitt.edu/~jdnorton/Goodies

Tutorial with lots of helpful diagrams, by physicist Tatsu Takeuchi:
www.phys.vt.edu/~takeuchi/relativity/notes

What you'd see if you fell into a black hole:
casa.colorado.edu/~ajsh/movies.html

Measure the speed of light by cooking marshmallows:
www.physics.umd.edu/ripe/icpe/newsletters/n34/marshmal.htm

About Quantum Theory

The best online introduction to quantum theory:
www.ipod.org.uk/reality

A do-it-yourself quantum experiment with a laser pointer:
www.tinyurl.com/2pguae

Movie demonstrating the strange way particles rotate:
www.evl.uic.edu/hypercomplex/html/dirac.html

About Cosmology

Cosmological tutorial, including fallacies:
www.astro.ucla.edu/~wright/cosmolog.htm

Frequently asked questions about dark matter:
cosmology.berkeley.edu/Education/FAQ/faq.html

Analysis of arguments against the big-bang theory:
www.physics.ucdavis.edu/Cosmology/albrecht/Myinfo/Burbidge%20Reply/Albrecht.htm

About Quantum Theories of Gravity

Video contest to explain string theory in 120 seconds:
www.discovermagazine.com/twominutesorless

Brian Greene's *Nova* series online:
www.pbs.org/wgbh/nova/elegant

About Extra Dimensions

A tutorial about extra dimensions:
indico.cern.ch/conferenceDisplay.py?confId=a02383

Musings about the fourth dimension:
www.tetraspace.alkaline.org/introduction.htm

Mac OS X program to visualize hypercubes and other shapes:
www.uoregon.edu/~koch/hypersolids/hypersolids.html

About Experiments

I won't bother listing individual experiments, since it's easier just to Google them.

The Large Hadron Collider facts and figures:
public.web.cern.ch/public/Content/Chapters/AskAnExpert/LHC-en.html

Animation of the LHC in action:
www.atlas.ch/multimedia/animation_lhc_event.html

Cool movie of graviton flying out of the brane in a particle collision:
www-cdf.fnal.gov/PES/kkgrav/kkgrav.html

Best portal site for dark-matter detectors:
lpsc.in2p3.fr/mayet/dm.php

Blogs

www.backreaction.blogspot.com. How do these physicists manage to have such fast but informed reactions to discoveries? They also do well-written primers on basic physics topics.

www.cosmicvariance.com. Must-read science blog, a joint effort of several of the rising stars in physics and astronomy.

www.math.columbia.edu/~woit/wordpress. Peter Woit's (in)famous Not Even Wrong blog critical of string theory. Never fails to be interesting, but the reasonable comments are hard to sift from the extraneous ones.

www.motls.blogspot.com. Luboš Motl's Reference Frame blog defending string theory. Sometimes so over-the-top that it's unintentionally funny. The truth presumably falls somewhere between his and Woit's remarks.

Index

Numbers

0-brane, 155
1-brane, 155
2-brane, 155
3-brane, 155
3-D space, 183-184
10-D space, 171-172

A

alternatives to string theory, 159-167
  "buckyspace," 163-164
  domino theory, 164-166
  loop quantum gravity theory
    atoms of space, 161-162
    challenges, 163
    loops, 160-161
angular momentum, 16
anthropic coincidences, 196-197
anthropic principle, 197-198
antimatter, Standard Model (particles), 96
antiparticles (fermions), 62
Apache Point Observatory, 274
atoms of space (loop quantum gravity theory), 161-162
Auger Observatory, 269

B

background-dependents (gravitons), 144-146
beauty, symmetry and, 211-213
beginning of time origins
  origins of time
    big bang, 121-122
    black holes scenario, 253
    ekpyrosis, 254-255
    gas of strings scenario, 252-253
    inflation, 251-252
    loop gravity and, 256
    overview, 249-250
    pre-big bang scenario, 253-254
Bell, John Stewart, 49
big bang
  cosmic expansion, 114-116
  cosmic inflation, 119-120
  meanings, 113-114
  overview, 111-112
  pre-big bang scenario, 253-254
  timeline of expansion, 116-118
  ultimate beginning issues, 121-122
black holes
  geography
    event horizons, 102-106
    singularity, 102-105
  Hawking Effect, 107-109
  information paradox, 109
  information storage capacity, 229-230
  loop gravity and, 245-246
  mini, 102
  no-hair theorem, 105
  origins of time, 253
  overview, 99-102
  quantum gravity, 110
  quantum theory and, 106-109
  stellar black holes, 101
  string theory and
    branes, 243-244
    duality concept, 241-242
    overview, 240-241
  supermassive, 101
Bohmian mechanics (quantum theory), 291
boson particles, 54-55

branes
black holes and, 243-244
D-branes, 155-156
types, 155
braneworlds, 175-176
brute facts, 191
bubble universe, 193-194
"buckyspace," 163-164, 207

C

Calabi-Yau shape, 180-182
causal sets, 165, 208
challenges
event horizon (black hole), 106
loop quantum gravity theory, 163
singularity (black hole), 105
spacetime, 89-92
Standard Model (particles), 92-96
antimatter, 96
hierarchy problems, 93-95
string theory critics, 277-285
time machines, 130-132
unification theory, 18-20
colors, fermions, 62
Compton wavelength phenomenon, 75-76
conservation of energy, 16
conservation of momentum, 16
contingents, 189-191
contradictions, theories of relativity, 30-32
coordinates, 34
cosmic coincidences, 124
cosmic expansion (big bang), 114-116
cosmic inflation (big bang), 119-120
cosmic rays studies, 269-270
cosmic string studies, 273
cosmological constants, 94-95
cosmology, 113
cosmos, comprehensibility of, 294
critics, 277-285
common arguments, 280-285
overview, 278-279

D

D-branes, 155-156
dark energy, 123-124, 258-259
dark matter, 122-123, 258-259, 272-273
decoherence, 56
deformed special relativity, 209
determinism, 45
dilaton, 151
dimensions (extra dimensions of space)
3-D space, 183-184
10-D space, 171-172
braneworlds, 175-176
Calabi-Yau shape, 180-182
small-dimension scenario, 178
sticky-dimension scenario, 178-179
visualizing, 173-175
domino theory, 164-166
doublets (fermions), 62
doubly special relativity, 209
duality concept
black holes and, 241-242
gravity/gauge duality, 232-233
grid of points view (spacetime), 202-205
T-duality, 203

E

$E=mc^2$ equation, 37-39
Einstein, Albert (theories of relativity), 27-39
contradictions, 30-32
general theory of relativity, 28
locality, 36
mass and energy equation, 37-39
principle of equivalence, 39-41
principle of relativity, 29
space and time union, 34-36
special theory of relativity, 28
speed of light, 32-34
ekpyrosis, 254-255
electromagnetism, 63, 79
electrons, 61

electroweak forces, 69
elementary particles, 60-61
emergence, 17, 225-227
energy
  dark energy, 258-259
  dark matter, 123-124
  mass and energy equation, 37-39
  negative energy (time machines), 129-130
  size and energy relationship (particles), 71-83
    Compton wavelength, 75-76
    forces, 80-81
    groupies, 76-80
    hierarchy of nature, 82-83
    manipulating particles, 73-75
    Planck scale, 81-82
entropy, 229
equivalence, principle of, 39-41
eternal inflation, 120
event horizons (black hole)
  challenges, 106
  overview, 102-105
events, 165
existence (origins of time)
  black holes scenario, 253
  ekpyrosis, 254-255
  gas of strings scenario, 252-253
  inflation, 251-252
  loop gravity and, 256
  overview, 249-250
  pre-big bang scenario, 253-254
experimental test categories, 262-264
extra dimensions of space, 267
  3-D space, 183-184
  10-D space, 171-172
  braneworld, 175-176
  Calabi-Yau shape, 180-182
  small-dimension scenario, 178
  sticky-dimension scenario, 178-179
  visualizing, 173-175

F

fermion particles, 54-55, 62-63
  antiparticles, 62
  colors, 62
  doublets, 62
  generations, 62
  quarks versus leptons, 62
fields
  Higgs field, 68
  particles and, 65
forces
  nuclear forces, 67-69
  particles, 64
    electromagnetism, 63
    size and energy relationship, 80-81
    weak nuclear force, 63
  scale-dependent, 80-81
fractal space, 207
frozen time problem, 91

G

galaxy, 113
gamma rays studies, 268
Gamma-Ray Large Area Space Telescope. *See* GLAST
gas of strings scenario (origins of time), 252-253
general theory of relativity, 28
generations (fermions), 62
geography, black holes
  event horizon, 102-106
  singularity, 102-105
GLAST (Gamma-Ray Large Area Space Telescope), 268
global symmetry, 214-215
goals, LHC (Large Hadron Collider), 266-268
grand unification scale, 78-79
grand unified theory. *See* GUT

gravitational waves (testing string theory)
 gravitational wave detectors, 271
 indirect, 270-271
gravitons
 background-dependent, 144-146
 overview, 138-139
 scale-dependent, 141-143
 spacetime and, 143-144
 spin, 140
gravity
 gravitons
  background-dependent, 144-146
  overview, 138-139
  scale-dependent, 141-143
  spacetime and, 143-144
  spin, 140
 loop quantum gravity theory, 7
  atoms of space, 161-162
  challenges, 163
  loops, 160-161
 principle of equivalence, 39-41
 quantum gravity
  black holes and, 110
  time machines and, 132-133
 quantum theory of gravity, 5
 strings and, 153-154
 studies, 274
 supergravity, 219
gravity/gauge duality, 232-233
grid of points view (spacetime)
 duality, 202-205
 noncommutative geometry, 205-206
groupies (particles), 76-80
GUT (grand unified theory), overview, 96-97

H

Hagedorn temperature, 240
Hawking Effect (black holes), 107-109
Heisenberg uncertainty principle, 73
hexacube, 174
hidden variables (quantum theory), 291
hierarchy of nature, 82-83
hierarchy problems, Standard Model (particles), 93-95
Higgs field, 68
Higgs particle, 266
history of string theory, 147-150
holographic principle, 228-233
hypercube, 174
hyperpyramid, 174
hypersphere, 174

I-J-K

implications, unification theory, 20-21
inflation, 120
 cosmic inflation (big bang), 119-120
 origins of time, 251-252
information
 black-hole information paradox, 109
 storage capacity
 black holes, 229-230
instruments, 9, 12
internal symmetry, 214

L

landscape (string landscape), 189
Large Hadron Collider. *See* LHC
leptons versus quarks, 62
LHC (Large Hadron Collider)
 goals, 266-268
 mechanics, 265-266
light (speed of), role in relativity theory, 32-34
local symmetry, 214-215
locality
 relativity theory, 36
 spacetime, 227-228
loop gravity
 black holes and, 245-246
 origins of time, 256
 smallness scale, 206

loop quantum gravity theory, 7
  atoms of space, 161-162
  challenges, 163
  loops, 160-161

M

MAGIC telescope, 268
manipulating particles, 73-75
many-worlds interpretation (parallel universes), 194-195
mass and energy equation, 37-39
mathematical universe, 195-196
matter
  dark matter, 122-123, 258-259, 272-273
  particles of, 62-63
microwave background, testing string theory, 270-271
mini black holes, 102
multiverse, 191

N

nature
  hierarchy of nature, 82-83
  symmetries of nature, 216-218
negative energy (time machines), 129-130
neutralino, 222
no-hair theorem, 105
noncommutative geometry, 205-206
nonlocality (spacetime), 233-234
nuclear force, 67-69
  strong, 63
  weak, 63

O

objective symmetry, 212
origins of time
  big bang, 121-122
  black holes scenario, 253
  ekpyrosis, 254-255
  gas of strings scenario, 252-253
  inflation, 251-252
  loop gravity and, 256
  overview, 249-250
  pre-big bang scenario, 253-254

P

parallel universes
  anthropic coincidences, 196-197
  anthropic principle, 197-198
  bubble universe, 193-194
  contingents, 189-191
  many-worlds interpretation, 194-195
  mathematical universe, 195-196
  multiverse, 191
  our universe, 191
  space beyond our horizon, 192
  string landscape, 188-189
  testing string theory, 274-275
particles
  Higgs particle, 266
  quantum theory
    bosons, 54-55
    decoherence, 56
    fermions, 54-55
    spin property, 56
    wave description, 52-54
  size and energy relationship, 71-83
    Compton wavelength, 75-76
    forces, 80-81
    groupies, 76-80
    hierarchy of nature, 82-83
    manipulating particles, 73-75
    Planck scale, 81-82
  Standard Model, 59
    challenges, 92-96
    electrons, 61
    elementary particles, 60-61
    fermions, 62-63
    fields, 65
    nuclear forces, 67-69
    particles of force, 63-64

photons, 61
virtual particles, 66-67
virtual particles, 77-78
philosophy, works in progress, 293-294
photons, 61
physics
tree of, 13-15
unification theory
challenges, 18-20
coexistence of science and religion, 22-23
implications, 20-21
overview, 15-18
shared efforts regarding, 23
Planck scale, 81-82
pre-big bang scenario, 253-254
principles
equivalence, 39-41
quantum theory, 45-47
relativity, 29
probabilities, shell game, 48-52
pros and cons, supersymmetry, 220-222
protons, decay, 272
proving theories, 264-265

Q

quantum gravity
black holes and, 110
time machines and, 132-133
quantum theory, 4-5
black holes and, 106-109
Bohmian mechanics, 291
decoherence, 56
hidden variables, 291
overview, 43-44
particles
bosons, 54-55
fermions, 54-55
spin property, 56
wave description, 52-54
principles, 45-47
shell game, 48-52
Standard Model (particles), challenges, 92-96
status of, 290-293
success of, 137-138
undocumented features, 56-57
versus relativity theory, 87-89
quarks versus leptons, 62

R

reductionism, 17
relativity theory, 27
contradictions, 30-32
general theory of relativity, 28
locality, 36
mass and energy equation, 37-39
principle of equivalence, 39-41
principle of relativity, 29
space and time union, 34-36
special theory of relativity, 28, 208-209
speed of light, 32-34
versus quantum theory, 87-89
religion, unification theory and, 22-23

S

scalar fields, 251
scale-dependent (gravity), 141-143
science, unification theory and, 22-23
self-explanatory symmetry, 212
shell game (quantum theory), 48-52
simple symmetry, 212
singularity (black hole)
challenges, 105
overview, 102-105
size and energy relationship (particles), 71-83
Compton wavelength, 75-76
forces, 80-81
groupies, 76-80
hierarchy of nature, 82-83
manipulating particles, 73-75
Planck scale, 81-82

size scales
  smallness
    "buckyspace," 207
    effects of spacetime foam, 199-202
    grid of points view, 202-206
    loop gravity, 206
  spacetime foam, 201-202
small-dimension scenario, 178
smallness scale
  "buckyspace," 207
  effects of spacetime foam, 199-202
  grid of points view, 202-206
  loop gravity, 206
space
  atoms of space (loop quantum gravity theory), 161-163
  "buckyspace," 163-164
  extra dimensions
    3-D space, 183-184
    10-D space, 171-172
    braneworld, 175-176
    Calabi-Yau shape, 180-182
    small-dimension scenario, 178
    sticky-dimension scenario, 178-179
    visualizing, 173-175
  space and time union (relativity theory), 34-36
  spacetime. *See* spacetime
space beyond our horizon universe, 192
spacetime
  "buckyspace" and, 207
  causal sets, 208
  challenges, 89-92
  effects of spacetime foam, 199-202
  emergence, 225-227
  gravitons and, 143-144
  grid of points view
    duality, 202-205
    noncommutative geometry, 205-206
  holographic principle, 228-233
  locality, 227-228
  loop gravity and, 206
  nonlocality, 233-234
  symmetry, 214
sparticles, 218, 267
special relativity theory, 28, 208-209
speed of light, role in relativity theory, 32-34
spin properties
  gravitons, 140
  particles, 56
Standard Model (particles), 59
  challenges, 92-96
    antimatter, 96
    hierarchy problems, 93-95
  electrons, 61
  elementary particles, 60-61
  fermions, 62-63
    antiparticles, 62
    colors, 62
    doublets, 62
    generations, 62
    quarks versus leptons, 62
  fields, 65
  nuclear forces, 67-69
  particles of force, 63-64
    electromagnetism, 63
    strong nuclear force, 63
    weak nuclear force, 63
  photons, 61
  virtual particles, 66-67
stellar black holes, 101
sticky-dimension scenario, 178-179
string landscape, 189
string theory
  alternatives, 159-167
    "buckyspace," 163-164
    domino theory, 164-166
    loop quantum gravity theory, 160-163
  black holes and
    branes, 243-244
    duality concept, 241-242
    overview, 240-241
  critics, 277-285
    common arguments, 280-285
    overview, 278-279

extra dimensions of space
3-D space, 183-184
10-D space, 171-172
braneworld, 175-176
Calabi-Yau shape, 180-182
small-dimension scenario, 178
sticky-dimension scenario, 178-179
visualizing, 173-175
history, 147-150
instruments, 9, 12
overview, 3, 7
principles, 8-9
testing
cosmic rays studies, 269-270
cosmic string studies, 273
dark matter studies, 272-273
experimental test categories, 262-264
gamma rays studies, 268
gravitational wave detectors, 271
gravity studies, 274
LHC (Large Hadron Collider), 265-268
microwave background and gravitational waves, 270-271
parallel universes, 274-275
proton decay, 272
proving, 264-265
time machines and, 246
versus loop quantum gravity, 7
works in progress, 287-289
strings
branes
D-brane, 155-156
types, 155
cosmic string studies, 273
defining characteristics, 150-151
gas of strings scenario (origins of time), 252-253
gravity and, 153-154
Hagedorn temperature, 240
inner life of, 152-153
strong nuclear forces, 63
Super-Kamiokande detector, 272
supergravity, 219
supermassive black holes, 101
supersymmetry
overview, 218-220
pros and cons, 220-222
symmetry
beauty and, 211-213
objective, 212
points of view, 222-223
self-explanatory, 212
simple, 212
supersymmetry
overview, 218-220
pros and cons, 220-222
symmetries of nature, 216-218
types, 213-215
global, 214-215
internal, 214
local, 214-215
spacetime, 214
unified, 212

T

T-duality, 203
testing string theory
cosmic rays studies, 269-270
cosmic string studies, 273
dark matter studies, 272-273
experimental test categories, 262-264
gamma rays studies, 268
gravitational wave detectors, 271
gravity studies, 274
LHC (Large Hadron Collider)
goals, 266-268
mechanics, 265-266
microwave background and gravitational waves, 270-271
parallel universes, 274-275
proton decay, 272
proving, 264-265

theories
  alternatives to string theory, 160-167
    "buckyspace" 163-164
    domino theory, 164-166
    loop quantum gravity theory, 160-163
  quantum theory
    decoherence, 56
    overview, 43-44
    particles, 52-56
    principles, 45-47
    shell game, 48-52
    undocumented features, 56-57
  relativity theory (Albert Einstein), 27-41
    contradictions, 30-32
    general theory of relativity, 28
    locality, 36
    mass and energy equation, 37-39
    principle of equivalence, 39-41
    principle of relativity, 29
    space and time union, 34-36
    special theory of relativity, 28
    speed of light, 32-34
  string theory
    alternatives, 159-167
    black holes and, 240-244
    critics, 277-285
    extra dimensions of space, 171-184
    history, 147-150
    instruments, 9, 12
    overview, 3, 7
    principles, 8-9
    testing, 262-275
    time machines and, 246
    versus loop quantum gravity, 7
    works in progress, 287-289
time
  frozen time problem, 91
  origins of time
    black holes scenario, 253
    ekpyrosis, 254-255
    gas of strings scenario, 252-253
    inflation, 251-252
    loop gravity and, 256
    overview, 249-250
    pre-big bang scenario, 253-254
  space and time union (relativity theory), 34-36
  spacetime, 34
    challenges, 89-92
    effects of spacetime foam, 199-202
    emergence, 225-227
    grid of points view, 202-206
    holographic principle, 228-233
    locality, 227-228
    nonlocality, 233-234
  ultimate beginning issues (big bang), 121-122
  understanding mysteries of, 289-290
time machines, 125
  challenges, 130-132
  negative energy needs, 129-130
  role of quantum gravity, 132-133
  string theory and, 246
  wormholes, 126-129
timeline of expansion (big bang), 116-118
tree of physics, 13-15
types of symmetry, 213-218
  global, 214-215
  internal, 214
  local, 214-215
  spacetime, 214
  symmetries of nature, 216-218

U–V

ultimate beginning issues (big bang), 121-122
undocumented features (quantum theory), 56-57
unification theory
  challenges, 18-20
  coexistence of science and religion, 22-23
  GUT (grand unified theory), 96-97
  implications, 20-21
  overview, 15-18

relativity versus quantum theory, 87-89
shared efforts regarding, 23
unified symmetry, 212
universality, 166
universe
multiverse, 191
our universe, 191
parallel universes
anthropic coincidences, 196-197
anthropic principle, 197-198
bubble universe, 193-194
contingents, 189-191
many-worlds interpretation, 194-195
mathematical universe, 195-196
space beyond our horizon, 192
string landscape, 188-189
uroboros, 200

virtual particles, 66-67, 77-78
visualizing extra dimensions, 173-175

W-X-Y-Z

waves
descriptions (quantum particles), 52-54
gravitational waves, 138
weak nuclear forces, 63, 67-69
worldsheets, 152
wormholes, 126-129